Landing Gently
by
Paul Smith

*Dedicated to My Darling Wife Monica
(to mark our Golden Wedding Anniversary)*

Published by Wolfgang Publishing
Wings, Court Street, Winsham, Chard, Somerset.

Printed by Rapid Print, Chard, Somerset.

Copyright Paul A. Smith B.A. Hons

ISBN 0 9524892 4 4

By the same Author   Whither Winsham 1992
               "Wars are wot you make 'em!".

Book illustrated with intertext B&W pen tinted drawings and maps.

Dealing with my early life on an isolated Somerset farm.
From May 1921 to 2nd Sept 1939.
Consisting of a coloured front cover, preface and 25 subject chapters.

Text 46300 words.
About 120 drawings.
Plus map.

Contents

Chapter			Drawing	Page
	Preface	*LANDING GENTLY* - Setting scene and introducing family		5
1	Last Load	Final journey to new farm. With Mother and twins.	13	9
2	Moving in	Moving into old farmhouse, unoccupied for 2 years.	14	17
3	Buildings	Study of farm buildings and their development.	15-16	22
4	The People	Study of local people who came to the farm	25	30
5	The livestock	An account of the farm domestic stock	25-27	35
6	Track Out	The quarried stone road that Dad built.	28	39
7	Fairyland	The magic of a small copse.	45	44
8	The Ram	Unravelling the mysteries of the water supply.	46	52
9	Around the Lanes	Early walks outside the farm boundary.	47	55
10	Education	School and beyond.	48	59
11	Paying Guests.	Mother takes in paying guests to keep us afloat.	65	72
12	Our Transport.	How we all get around ... other than on foot.	66	79
13	Milking	The mainstay of the farming.	67	84
14	Local Woodland	The wooded areas across the farm boundary.	68	91
15	'ITEM'	A small boy who caused chaos as a guest !	85	95
16	Hard Labour	I leave school at 14 to save the farm	86	98
17	Haymaking	The sweat and joys of the old traditional way.	87	103
18	Harvesting	The ways of the 20s.	88	112
19	Threshing	Steam power comes to the farm	106	116
20	The Poacher	A local 'fable. from whom I learn much !	107	120
21	Ferreting	Much harder than fishing	108	124
22	The Young Farmers	A social explosion for farm children.	125-127	131
23	Cider making	The making of real scrumpy.	128	134
24	The Wild Life	A short description and guided tour.	146-148	138
25	War Looms	I land myself in the Army for nearly seven years.	149	144

The coloured front cover depicts our arrival at the farm in 1921.
The back cover depicts my leaving the farm for the War on 2nd Sept 1939.

Each drawing augments part of the text and could be reduced to a thumbnail sketch at that point. Conversely it could be used to head a chapter or give a final flourish.

All the drawings have been done specially for the book with the exception of three I had published in 'The Young Farmer' magazine over sixty years ago.

The subject matter of many of the chapters has also been the subject of my articles, mostly some thirty years ago, in the Western Gazette, Pulman's Weekly News, The Smallholder and in broadcasts.

Preface
LANDING GENTLY

On the 17th November 1919 a Somerset farmer's wife had twins. They arrived just at dinner time, which must have been very inconvenient for the midwife. As they arrived just five minutes apart there must have been a certain amount of turmoil. Probably the kettle had to be reboiled or other panic operations put into place. It was indeed stated that the said farmer's wife's medical advisors were not even certain about it being twins at all.
It was also a fact that she had left the farm a few days before to travel to a small nursing home in the south coast town of Bournemouth. A wise precaution, no doubt, as the midwife person from the village back home also helped with the local lambing ... and had not enjoyed a good season ... quite a few lambs were ...er...lost.

This took place some eighty years ago and I was the younger of the twins .... by five minutes. Mother had said she was so shocked at the time that she called us by just one name each, Peter and Paul ... it could have been worse! Although we were not identical twins Mother dressed us alike and this later caused us great problems at school.

I personally recall nothing about the first two years of my life but have pieced together the events at that time from what I subsequently learnt from my parents. But this book does not start at the small fragmented farm around the centre of the village of Westhay on the Somerset 'Levels' where we lived at the time.

In early 1918 the British Army, decimated by casualties, were desperate for recruits. My Dad, who was running a small holding in the Village of Bawdrip - near Bridgwater - was called up but failed the medical. Being nearly thirty eight years of age at the time he was startled into accepting a possible worse fate and decided to get married. He was bred from a long line of small farming people and had left school at eleven to take over when his father suddenly died - it was said through a surfeit of 'scrumpy' cider from his own cherished orchard. I like to think that my grandad died with a broad smile on his face and a firkin of cider in one hand. Dad's mother and his two sisters never allowed him to do any domestic work at all.

Mother was the same age and was born in a large house outside the village. Her people ran a candle factory in Bridgwater. She had trained as a nurse in Bristol and Great Ormond St Hospital in London and had a piano playing qualification. She had spent all her adult life as a nursery governess to the very rich abroad in South America, India and with members of the French diplomatic service in Greece and Scandinavia.

Her prime duty was to take charge of one small - usually very spoilt - child. To teach her etiquette, English, French and the piano. As her employers were usually not English she taught them English also. Living with a large domestic staff she knew little of cooking, housework, and absolutely nothing about English farming.

So I can safely say that my parents never had a row in all their married life. They lived in different 'worlds'. For the first few years Mother had some help from a village girl who lived in. The effect was that Mother gave us twins the full 'treatment' to which she was accustomed and dad ruled his world outside. Knowing nothing about raising children he must have wondered how he would get any family help ...if ever.
The Great War ended after we were born and tenanted farms were being sold. Their fragmented holding around Westhay soon disappeared as the fields were sold. Mother heard on the grape vine that a farm was available in the hills near Chard - some twenty miles away ... as the crow flies.

The farm was one of a number of outlying properties that had been in an auction from the Lord Bridport's Cricket St Thomas Estate when it was taken over by the Fry family in 1919. The particular farm was not sold and all the buildings had been unoccupied for two years while the land was used by a neighbouring farmer to grow corn and graze sheep.

So Mother arranged a family mortgage and Dad bought the farm. Dad then managed in the spring of 1921 to move his cattle , pigs, two horses, poultry, two ferrets and two dogs with all the machinery, equipment and household goods to the new farm . He did this with roads at either end with the railway for the middle section. The problem for him must have been of almost biblical proportions. As Mother never helped on that move I never learnt the details. Mother had just packed up the smaller household supplies at the start.

The important point was that Mother had never seen the farm or even the house until she had made the final journey by passenger train to Chard Passenger Great Western rail station . She had a few boxes of household goods ,a large old double pram with us twins fastened in at either end and Judy , her spaniel dog.

Dad, dressed in his best boots and leggings met us outside the Station with horse and trap. Loaded us all aboard and , it being a very hot May day went up to the road junction in Fore Street were there was a water trough.

It is at this point that our story begins ......! It was sixty two years later that I finally left our farm for a retirement in the nearest village, Winsham.

Chapter 1
THE LAST LOAD

The whole Universe started with a Big Bang. ... although there was no one around to hear it at the time. but modern astronomers claim to still detect the distant echoed rumbling. Most local fireworks displays start with a big bang to rivet attention. But the bang that started my first infant memory was of a very different and more earthy kind.

It is fortunate that tiny babies do not recall the assault on their persons in the traditional task of 'bringing up the wind', Here after a lovely warm feed of breast milk they are often placed over the shoulder and given a few resounding thumps on the back. The resulting belch can bring a startled look to the face of the baby and a proud smile to the mum ... as if she had won a contest at Bisley!

Dad used to call indigestion " A touch of the gripes," But he only dealt with animal gripes and the animal that got the most frightening 'touch of the gripes' was Whinnie. She was the horse that was now in the shafts and when she got the gripes she not only groaned and stamped but tried to lie down and roll over on to her back.

She was the result of Dad using a 'thoroughbred' racing Stallion to 'cover' his old steady Shire mare called Flower. Whinnie seemed to have inherited many of the worst characteristics of both. She tended to have two speeds -walk and full gallop, with no reverse.

When Dad collected us from Chard Town Passenger Station it was the last journey of his momentous move. Although he had taken a churn of milk to the milk 'factory' on the way he was dressed in his best ' going out ' clothes with a waistcoat, best boots and leather leggings. But as it was not a Sunday he did not put on a collar and tie. Mother was in a long dress buttoned to the neck and a floppy straw hat kept on with a long hatpin. ...all in black. Still very much the governess.

It was a blazing hot May day and Whinnie was in quite a sweat with the flies continually worrying her so that she was constantly wagging her head, stamping her feet and swishing her tail. She was thirsty.

At the water trough she gulped down cool water and scowled at several other horses waiting their turn. Then holding the reins tightly Dad faced Whinnie down the hill to go out of the town and up the five miles to the hills at Windwhistle. Mother sat on the bench beside him with the pram amid the empty churn, several boxes and the dog on the floor behind.

Whinnie had barely gone a hundred yards when she stopped in her tracks. She was visibly expanding. She was shuddering with a fixed, intent look. Dad dropped the reins

"The GRIPES! " he shouted as he jumped out of the side of the trap and fishing against the tailboard produced a long pole. This he passed under Whinnie's belly and shouted to Mother, who had by then dismounted, to hold the other end. Massaging the belly by rubbing the pole back and forth it must have looked to the gathering spectators as if two people were balancing a horse on a stick.

After a short while Dad shouted "AWAY ...SHE'S BLOWING!" and deftly replacing the pole he regained his seat and gripped the reins. Mother was less fortunate and with her long skirt just managed to sprawl across the side of the seat with a cry of dismay that sent both of us twins yelling. Then Whinnie took control.

Horses cannot easily bring up wind forward but Whinnie suddenly gave an explosive snort and released the wind from her gut like a foghorn. Then amid a green haze she set off down the street and out of the town. Mother was heard to remark afterwards that Dad's furious driving was 'Like Jehu, the Son of Nimpshi.'
The iron clad wheels of the poorly sprung cart clanked over the loose gravel of the road and Whinnie's metal shoes clobbered like a kettle drum.

We soon sped past the houses and the road narrowed and meandered - between high banked hedges - up into the hills. Past the first milestone steam was rising from Whinnie, and as the sun poured relentlessly down, she slowed down to a walk. We got a wave from a road man who lived in the small thatched cottage beside the road and continuously trimmed, with sickle and guide-stick, the banks over his two mile stretch. It was also hot in the pram and we twins were only partly shaded under white hats with floppy brims.

Despite her blinkers Whinnie managed to grimace at a horse and trap coming in the opposite direction and to violently swerve when she spotted a sheet of white newspaper beside the road.

We also met a steam traction engine noisily pulling a trailer with cracked stones used for road repairs . With the narrow road Dad had to pull at the side and stop for mechanical traffic, as horses were only slowly getting accustomed to such modern methods of transport.

At the three milestone the verge widened and the gradient fell away. A gentle breeze indicated that we had reached the top of the hill. Dad suddenly shouted "Whoa!" and handed Mother the reins - which she reluctantly accepted. Before us the road was embraced by a high avenue of beech trees and above a flock of birds circled like a swarm of bees. The unmistakable sound of a rookery wafted towards us

Dad went to the back of the trap and produced an enormous black umbrella, which he called a 'gig'. Unfurled, its stem fitted into a socket along the bench seat. The sky was still blue with a few wisps of high cloud.

As we moved again and approached the avenue the road was spotted white as if it had recently snowed. Under the avenue of trees the reason for our upper protection was plainly obvious. The rooks were very busy above rearing their chicks in the high nests and they obviously resented the traffic in the road beneath. The resulting shower plopped noisily upon the gig and even Whinnie was becoming speckled with white spots.

Further along the great avenue of trees parted along one side and the great shining beech boles gave way to a long thatched building with a swinging sign ... THE WINDWHISTLE INN. Locally known as the 'Wustle'. At the four milestone there was a great gate beside a stone built lodge. Across the road was an enormous long pile of flint stones dwarfing the small man - with heavy metal gauze spectacles - who sat on a three legged stool and with a little hammer was cracking each stone. He was the County Council official stone cracker. Whinnie just snorted at him as she passed.
Just past this point the high banks had gone and great vistas of the Somerset countryside stretched out to our left in greens, blues and distant purples. The canopy was replaced by clumps of great towering beech trees in the fields. To the right there were single trees of sycamore, silver pine and holly.
Whinnie was now reduced to a tardy trot as under a massive spreading beech on our right was an old oak gate. This was our entrance to the farm .... with the buildings far down out of sight across two fields.
As Dad held open the gate Mother, very gingerly, guided us through . There was now a strange silence as the iron wheels went across the field to the far gate. Inside was Flower, glad to see her one-time foal. The field was very steep and we looked down over the farmstead - the farmhouse and surrounding buildings and trees - at the bottom. Beyond was a deep valley backed by high woodland.

It was all Mother could do to stay in her seat down the steepest part of the hill. We twins could even sense that something was wrong as we cringed deep in the pram as it lent on its two side wheels. Whinnie tried to turn her blinkered head to view our discomfort. It was probable that she intended to race Flower to the bottom. Leaning back with the reins tight Dad managed to restrain her to some degree But we still unceremoniously arrived at the bottom of the hill at a frightening angle with Whinnie sitting back on the breeching with all four feet held straight out in front .... skidding for the final fifty yards.

In front of us was the gate to the barton, the hay rick area with a stone wall surround. Mother was too frightened and exhausted to notice what looked like birds upon the buildings and up in the trees. On closer inspection they were seen to be hens. Another of Dad's special breeds. They were specially bred Anconas that were 'fox-proof' as they lived mostly above ground level.

Landing Gently — 13 — by Paul Smith

LANDING GENTLY by PAUL SMITH

Chapter 2
Moving In

I vividly remember that Mother was a stickler for cleanliness and good order. So what she found when she opened the back door of New House Farm must have shaken her to her roots. She sat down on a box and, clutching us twins with both arms, burst into tears ... only for a few moments, of course, for Dad was busy out on the farm and the British (even servants) - in far flung places around the World must always show a stiff upper lip. In farming society the house was woman's work so Dad would have not understood. .

In his efforts with the move Dad had pushed and shoved all the household goods into the back kitchen . As this room was a lean-to with exposed rafters and the farmhouse had not been occupied for nearly two years - and all the main farm buildings were thatched - everything was heavily draped in a thick veil of cobwebs. Thousands of eight-eyed monsters with black legs stared out upon the new intruders. Some of the cobwebs formed a curtain in front of the next door.
Soon Mother was searching around for some sort of brush when Dad turned up with his own version of an anti-cobweb device. This contribution to easing our plight consisted of holly twigs tied with string to the ends of long hazel sticks.
With these Mother managed to wrap bundles of cobwebs into tight balls. So that now angry spiders, deprived of their homes, sought refuge amongst the assembled furniture. Mother was not entirely convinced that the larger specimens were not poisonous tarantulas.

To add to our discomfort this room had a strong smell of very stale cheese. For the shelves along one wall sprouted greenish moulds and old cheese cloths. Flies buzzed around to fatten up the predatory spiders. .

The next room was called the pantry. It had one cold tap for water and what was presumably meant to be a safe. This was a large wooden box with metal panels to ventilate it and keep the insect life at bay. The many metal patches secured by tin-tacks were covering holes made in the woodwork by mice. The spider population in this room were mostly confined to the ceiling. Mother's long dark hair was carried in a 'bun' but she had by now donned a large brimmed black hat secured by a long hat pin. The cobweb adorning this pin made is resemble an ostrich feather. We were clad in identical short trouser suits with our white sunhats which were also adorned with wisps of cobweb.

On through the next door was the living room. It had two large oak beams and in a large chimney cavity resided a rusty old kitchen range. Like the rest of the house it looked as if the previous residents had left in a hurry for the ashes had spilled out on to the, very uneven, flagstone floor. A few logs and some sticks had collected a covering of dust and a small pile of soot rested in the chimney breast.

Clearing a small patch with a brush Mother had taken our double pram with us inside into this room as it was the only comparatively clean spot for us all to eat. The room was dark even on a sunny May afternoon as the mullioned window was small and grimy.
In, no doubt, a spirit of adventure Mother lit the fire with the available sticks and logs lying around. Soon a party of mice and a large toad decided it was time to leave.
With the fire established Mother had to ground us again so that we could resume the safari. Like a hen with chickens we followed close to her voluminous skirt .... she being too fearful to leave us alone with the fire.

So to the next room this time again with bare rafters and tiles, containing the water tank on a frame in the roof. The flag-stone floor had a section that was covered with boards. This to be a forbidden area as under those boards was the old 60ft well into which a hand pump was once piped. I subsequently discovered that if a small stone was pushed through a crack in the boards there was silence for a few seconds and then an echoing distant splash.

Back again to see if the fire had properly 'caught' mother drew a kettle of water from an outside tap and put it on the hob. The whole house was beginning to smell of wood ash as we went out of another door into a darkened hall ...there were no windows here and the front door was tightly closed. Another door led into the sitting room and Mother found that it looked out into the supposed garden. Apparently as gardens were not in favour with our predecessors they had removed the fence ... or it could have fallen down ... as cattle had obviously been 'roosting ' amongst the abundant growth of nettles and thistles.

This, we discovered, was more than a little unfortunate as the path outside led along the wall to a little green door.. This was, it transpired to be another 'forbidden' area. This was the only 'privy' in the house and the hole in the boarding was large enough for a small child to fall through and land in the open cess pit six feet below. Mother said that chamber pots would be sited at vantage points inside the house.

Mother opened the sitting room window to 'air' the room ... just in time to catch sight of Dad racing past after a bullock and screaming to his two dogs to head it off.. A few moments later he raced back in the opposite direction and Mother shouted hopefully "Tea ready shortly."

Back in the hall Mother prised open the front door to reveal what 'beasties' were in residence. Another shock ... the floor led back to a far door and the front stairs, but divided so that part led directly down steps into the 'bowels of the earth'- to the cellar.
From this emanated a stale smell of overripe mushrooms, stale earth and rotting potatoes.

Guided by Mother, as there was no guard rail, we went on - past the narrow winding unlit stairs leading to the floors above - to the far door. Mother was beginning to show signs of anxiety as perhaps the though had occured to her that some of our predecessors had perhaps perished and we might soon stumble on their bodies. This door led down a step into the 'bathroom'

There was a cold tap in the corner at the foot of a large enamelled iron bath, Across the concrete floor was a 'copper' under a wooden lid . Remnants of a fire beneath with a pile of ashes on the floor indicated that coal as well as wood had been used.. The chimney outside sprouting out of a slated roof was well positioned to shower sparks on the thatch in an Easterly gale. The bathroom had three doors none appeared to have locks on them.
The middle door led down a steep step into an earthen floored room , called the coal cellar, Here was a raftered ceiling and wherein dwelt a fine collection of spiders , small piles of coal dust and rotten wood all tied up with cobwebs. Although it had no window - and a door led directly to the cattle yard - it was, in the fullness of time - to be used to house a couple of cider barrels, roosting turkeys and firewood ...also sundry dumped discarded furniture. At the far end of the bathroom the door led back again to the outer kitchen. The ground floor route was thus circular .....a delight to two small boys in years to come.

As Mother prized open this last door she peered with horror onto a litter of piglets who had just entered amid the pile of furniture and boxes. She rushed to the window that looked out onto the yard and shouted "Lan!"

At this point Dad arrived at the outer door carrying a token gift of a dead rabbit and a clutch of eggs in his cap .... I was to learn that he always carried eggs in his hat. Whether he liked to live dangerously or just found it cheaper than buying a basket ... I never found out. Seeing the pigs he called up his trusty 'hounds' whose efforts managed to substantially rearrange the furniture and boxes before they and the piglets - covered in cobwebs dashed into the yard to rejoin their - by now very angry ... mother sow.

While we twins were just called Peter and Paul in a hurry our parents were obviously named at leisure. Mother was called Bea - short for Beautrice - a name that suited her. Dad called her Mother to her face but, as most farmers, used to refer to her as 'The Missus' when speaking to others.

Dad was known as Lan, short for Orlando, although I later thought he should have been called 'Land' .... he could not obviously have been called 'soil' 'Earth' .... after his sole interest.

It was at this point that we had our first meal. .. a sort of late tea. Mother was very much in charge and - even in the prevailing chaos and grime - required Dad to remove his boots outside the kitchen door. She found a table cloth and even laid the table so we all could 'sit up' properly. She would sometimes overlook the fact that Dad had not washed his hands before meals ... one of the features of Windwhistle Farm was a shortage of water. He had after all wiped his hands very thoroughly on an old sack.

The chief topic of conversation at that meal was the extra help - both on the farm and in the house that was expected to arrive in the morning. Mother soon cleared the table and Dad - rebooted - disappeared into his own world. Mother gathered us twins together and armed with a holly stick and a lighted candle, resumed the safari. This time from the bottom of the windowless stairs at the end of the hall.

The candlelight threw eerie moving shadows on the walls and there were three doors into bedrooms which Mother opened to let in some light ... so that she could release her tight grip on us. There was nothing particularly dangerous in this part of the house although Mother was a bit dubious about the bats that occupied several dark corners. Were there vampires in rural Somerset?

Next was the creaking door to the 'down the stairs' bedroom over the coal shed that looked out on the farmyard. Here we could see how the main buildings -like the house had two storeys and were thatched with walls of local flints and sandstones. From these main buildings came lean-to sheds with heavy tiled roofs and timber fronts. With the house they made up a central farmyard. Great trees watched over the whole. From the windows we could see no other houses or even roads ... we were very much on our own.

A fourth door on the dark passageway led to the fourth bedroom over the living room. Despite the spiders and dust Dad had managed to get a double bed and our cot into this room ... which overlooked the rick yard. So on into the darkness of the steep back stairs and a brief look into the fifth bedroom. These stairs would also prove fun for small boys!

Although Mother must have been relieved not to have found any bodies she must have been alarmed at the wide windowsills ... due to the thick stone walls. If a small child could get upon the sill could it then easily fall out of an open window.

That evening - when Dad was hand milking the small herd of cows Mother took us, and this time with Judy her Spaniel on a lead - on a quick look around outside the house.

The farmstead was indeed isolated with not even a real track across several fields to a public road. There was a group of implements scattered around the rickyard. A few hens, ducks and gleanies (guinea fowl) busied themselves around and much to Mother's consternation they seemed to expect her to throw them some grain and we twins had to dodge them to avoid getting pecked ....in a friendly way of course. Mother let us peep inside the barn but we dare not enter the cattleyard as a cow was loose there with her calf.

It had indeed been a long day for us all and Mother - after a judicious bit of dusting put us to sleep in the cot upstairs. I now know that Mother and Dad worked until darkness and beyond. Tomorrow some help would arrive.

Chapter 3
AROUND THE BUILDINGS

As a small child I always loved nooks and crannies. They were sort of my size and the old farm buildings - thick stone walls, timbered roofs covered by straw thatch - were full of them. Especially in the winter months when hay, straw, rootcrops or sheaves filled the floor spaces so that I could clamber up under the roofs.
I was not alone for up there were a good selection of spiders, beetles and the odd sleeping butterfly. The farm cats also favoured these places to rest or even have their kittens. Often I would see mice or rats scurrying along the tops of the walls.

The barn was partly given over to storing mangel worzels which had been thrown in from the north facing high door and retained by a high wooden partition. The centre section of the barn contained the brightly coloured mangel slicer which was operated by a large hand wheel and the slices fed to the cattle from a large wickerwork 'Willy Basket'. The barn centre also contained slabs of cotton cake and a hand operated 'cake crusher'. There was a chaff cutter to cut up hay for the horses to eat from nosebags at work. Dad also kept hessian sacks of pig, poultry and calf food.

There was a home made strawrake for combing out the reed for thatching, from the sheaves which filled the far end of the barn. There were two narrow slits in the wall - like arrow slits in a castle - to look out upon the rickyard beyond. The large double doors had a small exit door within one side.

The opposite side of the barn had a narrow high gate leading to the cowstalls. There were fifteen tie-ups behind a narrow forebay in front of the cribs. The stalls had a right angled bend. At the far end was a double bay for calf rearing. At this end behind a strong partition was the tie-up for the bull. Under this stall roof, on the rafters, was a favourite annual nesting place for swallows. This stall had a corrugated tile roof.

Facing the stalls across the yard was the stable block. Here was stabling for four horses with a slim bay with racks .... dripping with harness and chains. Above the horses , entered by a vertical ladder fixed to the wall was the 'talet'. An outside wooden door allowed hay to be passed in from the field and the floor ended in racks over the horse bays. When the tallet was full it was a real ,cubby hole, The roof often had sleeping bats at the apex and outside, under the thatch overhang, was a favourite site for the house martin nests. The odd one with the enlarged entrance hole having been taken by a family of house sparrows.

Along the buildings was an open fronted section with a ladder to the square 'room' above. This was called 'the granary' and had a smooth floor and smooth plastered walls. There was no sign of a gantry for a pulley when we came to the farm. This room had no obvious use as it was not possible to carry a sack through the entry hole in the floor. The sparrows took full advantage of this and nested in the inner ends of the roof thatch.

There was a lean-to shed on the end of this building used variously for ducks, pigs or even calves. This had a rough thatched roof. Beyond this was a long tiled pig-sty block with a properly timbered set of runs. The floor had - so Dad related - been properly concreted once. But the pigs had been put in before the concrete was properly set. It looked as if they had tried to eat it. So mucking out here was almost impossible. Blue-tits and wrens usually found somewhere to nest in the poorly pointed walls. Behind the pig unit were three gaunt tall hornbeam trees which touched the back walls and were very gradually moving the stones in the wall.

It was my early interest in the buildings that divided Peter and me. Despite Mother's disapproval at my getting dirty, dusty and gatherings cobwebs about my person Dad said I was alright. Although he appeared far too busy to keep an eye on me. I loved the freedom to creep around the building ... except at specially 'forbidden' times. These I soon discovered were when the male animals were loose for mating purposes ... often a spectacular performance not for children's eyes. .

This was also especially true when Dad was constructing the 'SHED'. This was no small building, as it was some twenty feet high and forty feet square. As with a projected farm road it was to cost practically nothing. To Dad it was rather like the Acropolis was to the ancient Greeks.
The construction of the shed was thus somewhat shrouded in mystery. For not only to help make it but then to stand near and watch Dad swinging about like an Orang Utan was potentially dangerous. Even in those far off unenlightened days an employer could not ask a man to commit possible suicide as it was built with no scaffolding. Mother would divert her gaze when looking out of windows as she did not wish to see Dad crash to earth. Miraculously he survived.

The Shed basically consisted of fir poles, culled from around the farm, sunk in the ground at ten foot intervals around the perimeter ... and a few in the centre. these connected at the top by cross-pieces on which was nailed, or wired, galvanised sheeting. There was only a slight pitch to the south but even then the wind would tend to lift the whole structure up a few inches like a giant aerofoil. Dad did try putting heavy stones on the top but these had an awkward habit of rolling off and bouncing away at awkward moments as lethal weapons.

To lessen the wind effect he sheeted the top end and the lower sides he filled with packed faggots of brushwood secured with wire. There was an open gap left top and bottom.

When the Shed was finished Dad could leave a wagon loaded with hay or sheaves inside as required. He then gathered up his old carts -each with something broken or missing; a second wagon with a cracked shaft and loose wheel; his grass mower; the trapper; the swath turner and other numerous oddments of possible agricultural usage. Most had been fished out from overgrown brambles and nettles around the farmstead. All these he packed into the depths of the shed.

He now nailed hen nest boxes to every post. Then the growing flock of miscellaneous poultry were banished from outer sheds and the barn end and encouraged to roost in the shed. There were the gleanies - who had been chased out of the Lilac bush in the garden.. Some of the more sedate Anacona hens that had come down from the roofs and high trees were persuaded to join them. Finally there was a cockerel or two who could not be caught for the table, two bantams who lived in a nearby holly bush at night who had to be flushed.

The result was all the machinery soon gained a patina of guano and Mother counselled us not to play in the shed ... as the patina had a strong stink.

At that time - in the mid-twenties - the livestock on the farm was increasing and Dad's considered success with the Shed encouraged him to build his second 'no cost monument. This was the 'piggery'.

The breed of pig we kept was the Wessex Saddleback. This was a grazing pig that ate grass, clover, most soft weeds and - if not frequently 'ringed'- dug up the fields as well. Dad had decided to have a field devoted to the pigs; the 'Pig Field'. He had discovered that there was a patch in the middle of a field towards the Purtington road that had a platform of sandstone rock a few inches below the surface. This was to be the floor of his piggery.

MOTHER AND CHILD

Landing Gently   by Paul Smith

"I DECLARE THIS ROAD OPEN!!"

FROM THE CHARD.   FROM THE CHEST.

ONE MORE FOR THE ROAD

This time Dad constructed a square edifice with four individual litter pens on one side and a couple of larger pens on the far side. Again with galvanised but now with an apex roof sloping on both sides. Apparently dad had not invented guttering. The walls were made of brushwood faggots held behind rough coppice timbers.

He made the gates of old boarding with strips of odd sheeting nailed on to prevent rooting by adult pigs. Straw or dried nettles or even bracken served as bedding.
All this made for very cosy living quarters for pigs and also an increasing colony of rats. They lived in the bases of the walls and even found a place to dig some tunnels in the earth. As they were about thirty yards from the boundary hedge they made a straight tunnel to the hedge literally in the roots of the grass. They could actually be seen inside this tunnel.

This was their undoing. Much to Mother's initial horror I befriended one of Dad's ferrets. She was a small rather vicious strain. She loved attacking rats and was quite fearless. I soon managed to tame her and she would sit in the pocket of my 'outdoor' jacket. When the pigs were out grazing I would let her loose in the piggery where the rats would soon flee in terror. Sitting cross-legged by the long run I would swat the rats with a short thick stick as they pushed their way to the hedge. Often the pigs would happily eat the carcasses when I had left.

Dad soon discovered that to carry out water and food to the piggery with his four-bucket yoke was too tiring. So he decided to bring the pigs nearer to a place just below the garden for feeding and watering. This without the pigs getting into the steep part of Tip Hill. It would require a new fence down the field.
To this end he invented his 'no-cost' fence. To supply the material for this he had the two men and himself cutting brambles by the fork full and placing these on a row of short stakes he had hammered into the ground acrossthe field. Thus was made 'The Inner Pig Field'.

All this took place towards the latter half of the 1920s and the first tremor of the great depression to come was felt. Apart from a small lean-to shed at the back of the stable wall in Home Field the days of Dad's expansion were over.

Chapter 4
THE PEOPLE

On our isolated farm, with no other dwellings in sight, I recall various people turning up in the daytime. They must have walked from the settlements a mile on either side. They seemed to arrive very often wet, windblown or exhausted. At first the men or boys helped Dad out on the farm but Mother also had some help.
When she was in India, she used to tell us, a local woman or girl used to be available to mind the child in spare moments when she was busy. They were called 'ayahs'

Mother told us that at our first home, a village girl used to act in the same capacity - but we were then tiny and the house was in the middle of the village. Mother, I recall used to praise this girl, she was supposed to move with us but she committed the unforgivable sin .... she died.. Mother did not seem to blame us. Our new Ayah seemed fit enough to walk over most days from Purtington and her chief task was , of course, to keep us together. We called her Betty .... and we liked her very much. She was jolly and, we thought, seemed unlikely to die. She promised us not to die.

She was expected to keep us happy with the aid of a pack of 'happy family' playing cards, a teddy bear, a 'boy' doll and a stuffed toy green parrot. The real problem was that we did not have two of anything, except perhaps two soft balls which were normally hidden away. Mother used to tell us not to vent our anger on Betty but only on one another. or not at all. I still wonder why our previous Ayah called Olive, died.

One of Betty's other chief duties was to .see to our excretory needs. To this end Mother had positioned a series of chamber pots (we called them 'articles') throughout the house. We were not allowed to go near the garden privy, as not only might we throw our toys down the hole into the deep cess-pit four feet below, but we

well might actually fall through ourselves. So Betty had to empty the pots as required..
Betty had to train us to piss in these pots while standing and we thought it strange why she could not demonstrate the technique herself. So after we had finished she used to squat over a pot and we stood with her listening to the 'music'. Our infant giggles must have been the first arousal of sexual instincts..

About three times a week Mother used to have 'a woman ' in to help in the house ... ... in fact do all the cleaning and scrubbing. She was a widow known just as Mrs 'M'.
She always seemed very old with her long black dress and shawl . She also wore black boots with galoshes for travelling across the fields. She had to be resuscitated on arrival with a cup of tea and a few biscuits.

Mrs 'M' was almost spiral in shape - about five feet in all directions. Her slowly progressing work was liberally punctuated by initiating Mother into the finer points of local gossip ... much of which Mother probably hoped was not true.
When she would first spot us each day in front of Mother she would remark - in her rather squeaky very local accent "Ha, Missus! bain't 'em .ever growing?" As if we were kittens or chickens.
Although her accent normally missed out all the 'h's in speech .. just for Mother she would put them back elsewhere in the sentence Thus two of her favourite observations would be,
    "They howls be a 'ooting! . " or "They 'eifers be a heating they turmuts !".
.
It gradually transpired that practically all the local people were related to one another. and many had never been further than one local town .. either Chard, Crewkerne or Ilminster. Although Dad spoke with a Somerset dialect his tone was milder than the more guttural voices of those further South, near the Dorset Border It was also quite distinct from the Devon accent barely ten miles to the West.

Within a year there were two important changes in the house. Dad had found some boards around the buildings and constructed us a new (no cost) 'privy' in the back kitchen, Inside was a modern ELSAN toilet.
The other development was of a much more 'spiritual' nature. Mother had a piano delivered which was to reside against the wall near the kitchen range. On it were two large folders of sheet music. Strauss Waltzes and a complete set of the Piano Sonatas of Beethoven. I do not know whether this early form of 'Music While you Work' was exactly appreciated by Mrs M or even Betty but this early 'music at the breast' -as it were - set my (and Peter's) musical tastes for life.

Outside, on the farm , Dad had his own helpers. There was , at first, a solemn elderly man who used to deal with the horses ..Mother told us his name was Ned and he was the carter. He used to spend much time in the stable and cussed the horses, especially Whinnie, if they failed to obey his commands. But after a time he seemed not to be there and Mother said he had died. Strangely , in modern times, we were not surprised for we had old parents - at just over forty- and sixty usually meant wearing black, having some form of limp or stoop and awaiting the end.. Betty had promised us not to die ....at least not yet. .

Many of the young men from the villages now had their names on the War Memorials but there were a few left. The first man who I recall was called Joe - in fact a lot of young men were called Joe and most had biblical names- like Emanuel , Enock, Joshua or John ... but never Matthew as this was a local surname.

The first Joe was not a favourite with Mother This was because she always insisted that we twins both wore white sun hats when we went out with her pushing the pram. Neither of us liked sun hats and Mother had to distract our attentions while she slipped the hats on our heads .. a tricky business. . We also had a slight aversion to woolly coats.

Despite being warned about it, Joe would see us going out and - apparently twins being rare in villages at the time - come up to admire us and say something about our 'lovely 'ats misses'. Like greased lightning two white hats and often a woolly coat or two would leave the pram aimed at the nearest puddle or mudpatch, Mother was not amused.

About two years later the second Joe, arrived who helped with the heavier work mucking out and stock feeding who Mother allowed us to follow around as he seemed to like small children. We soon noticed that he had a habit of suddenly stopping his work and making nasty faces. We thought this was a joke until Mother, who knew a lot about the War told us the truth.

Many of the Army survivors were ' full of the metal'. The German guns would fire shrapnel shells that exploded above the ground filling the air with tiny, jagged bits of red hot metal. Many men were killed but for the survivors there was no way that all the pieces could be extracted so the metal stayed in the body catching in the muscles and other moving organs. After a few years Joe did not appear ... perhaps Mother kindly did not tell us the reason.

During these early years in the Spring I recall that Dad 'put on' a tramp. His name was Absalum, and he would just turn up and Dad would let him usually cut thistles, do a bit of hoeing or even do some mucking out. To me he was very, very old but very difficult to actually see. His entire head and face was obscured by a mass of whitish mangled hair which reached almost to his middle. The only visible garments were a very long blackish coat and a very dirty cap. In damp weather he had an old sack over his shoulders and he wore a red handkerchief - or what had been a red handkerchief. - around his neck. Dad let him live on a pick-full of straw in a corner of the barn for sleeping and Mother used to put some hot tea in an old metal pot which was about his only possession. I never heard him speak but I was once frightened when I heard him humming to himself

Mother told us he lived on raw eggs, a drop of milk, raw mushrooms and raw turnips. Then one day, perhaps up to a month later Dad would find that he had gone. After several years he did not turn up ... Mother thought he might have died .... like everyone else!

Chillington - a mile over the hill - was a rich source of workers. It was a change to see young men coming on who had not been muted by the Great War. Enock and Emanuel - otherwise known as Nocker and Manker - were two brothers in their teens who came together and were the first people with rich local accents that we heard talking together. For all the others said very little to either Dad or Mother within our hearing . For in this isolated self contained population we , including Dad , were regarded as 'voreigners'.

Dad usually let them work together around the buildings ... feeding animals or doing the mucking out. I was even allowed to try to help at times. Peter after a few years was beginning to spurn the idea of going out on the farm ... even at that age he preferred to look at books, listen to mother's wind-up gramophone or even listen to a little Beethoven on the piano. He was becoming a mother's boy.

So I would be allowed to watch Nocker and Manker working . Listening to their conversation was almost like hearing another language and I am sure that they realised that I could not understand much of what they said. I soon gathered that they did not exactly find too much joy in their work My inability to repeat their phraseology may have been just as well. Mother would not have approved ...neither would Judy , our spaniel house dog. !

All animals were, I gathered, often referred to as 'buggers' The larger ones as 'girt buggers' and the smaller ones as 'tiddy buggers'.

Listening carefully I also gleaned that all creatures were referred to as 'ee. As in 'Ee be goin' out to grass..' (When referring to a cow) This appeared to be different with the bull Here 'ee' then became an 'er.' But all creatures that caused trouble could be referred to as 'Thic bugger.'Mother had been very careful to make us twins talk distinctly, keeping each word separate. So it took me some time to fathom the meanings of such short phrases as :-
' No ... dissn' (no you did not) . 'Cassn wossn ' (can you not or would you not) or such a piece as "Dissn thee bin t' zeed 'un 'oller " (Did you not go to see him

LANDING GENTLY   *by* PAUL SMITH

shout.) when Manker went out in the yard to see if Dad was shouting for them. I also learnt words like that 'baint' meant 'is not', dumpsy was early evening, a nestletripe was the small one of a litter, and that 'zuent' meant evenly distributed.

In due course they left and their place was taken by Art. He was an older and very jovial character who used to play football with us in his dinner hour. He was notable because he introduced us to keeping homing pigeons. Dad always kept a pair as messengers on our car journeys but Art took us to Higher Chillington where several men bred them. Dad actually rigged us up some nesting shelves in the old garden privy - a catching grid on the window and the old seat securely boarded up. Art took the two of us to Chillington to a row of simple houses in a place called 'Two Gun Battery' and another row given an equally obscure military title.

Here we saw a flock of pigeons. There were blue checkers, red checkers and even pie-bald and almost black specimens. I recall Art telling us that some breeders sold older pigeons who would return to their old home, after of course they had been paid for, Peter and I bought a pair of baby red checkers who had not flown outside.

This time was in the early thirties and farming was going into economic depression ... with apparently the rest of the world. Art left to get married and live away elsewhere. Chillington had no piped water and there was only one tap at the side of a lower road.

For a few years Dad employed a young girl called Florrie who had left school and had learnt to hand milk cows. and was the daughter of the Mrs M who helped Mother many years previously. I would help her doing such jobs as collecting eggs around the buildings and feeding the poultry. She too left to get married.

We then had Ernie, an older man who arrived on a bicycle each day By this time there was a family living in a new house and petrol filling station along the main road called 'Windwhistle Garage'. They had two boys and a girl our age. In Purtington a family at Lower Purtington Farm used to invite us over to a field to play cricket in the evenings.

It was 1935 when the farm reached a low ebb in economics. Ernie had to leave and I too had to suddenly leave school on my fourteenth birthday to take his place. I would have no direct pay. This took me up to the first day of the next War.

Chapter 5
THE LIVESTOCK

Mother's realm was very definitely inside the house. As soon as we put a foot outside a door we entered a different world. This was Dad's realm, occupied by all sorts and sizes of farm creatures; some of which seemed to move at great speed followed by two barking dogs and Dad, with a big stick, in hot pursuit.
The whole place to us was teeming with feathered or hairy stock who were mostly larger and not entirely friendly to 'man cubs'!, So Dad had decreed that, from an early age, we carried a stick to defend ourselves. To this end a few suitable knobbly sticks stood at attention inside all exit doors.

Carrying sticks had two other effects, firstly it prevented us making real friends with the livestock and secondly we could act as a possible barrier of waving sticks to help Dad guide stock the way he wished them to go. This required some nerve as Mother would not welcome us being run over by sundry livestock.

The main area at one end of the stack yard was where grain was scattered for the, very free range, poultry. There was great variety with a few turkeys, some geese, ducks - usually podgy Aylesbury's and khaki campbells or even Indian runners. Guinea fowl - we called then 'gleanies' constantly saying 'go back, go back' as they circled the flock with wings ready to take flight. The fowls were equally variable. It seemed that Dad had randomly collected odd birds. There were Barnavelder and Cochin types. Rhode Island Red and Light Sussex. Dad's special breed were Anconas - a light breed that often flew like birds and kept out of the way of foxes. There were always several cockerels. usually awaiting slaughter for the table. But always a few to ensure that any eggs not found in a nest would ultimately be brooded and bring off chicks if the hen became broody..
The problem arose with collecting eggs. Whilst a few proper nest boxes were suitably positioned no one seemed to have told the poultry - who much preferred to chose their own nests, usually tucked away in odd unused corners , in deep sting-

ing nettles or - in the case of the Anconas - amongst roof rafters, on roof ridges or high trees. A long ladder was kept available. We sometimes met eggs that must have remained undetected for many months. These were not only addled but would explode if shaken ... the smell was of course of 'rotten eggs.'

A second problem was getting poultry to come inside a building at night. The local foxes were very cunning and despite the dogs being tied with string at the end of long chains outside the doorways near their kennels, the foxes would still get poultry in the night. It was said that a fox walking in circles under a roosting fowl would cause it to become giddy and fall off its perch.

So that, from our bedroom we would hear a squalking fowl ... a scurry of barking dogs as they strained and broke the string that secured their collars to the chains ... then Dad rushing down the stairs . He carried his shot gun and a hurricane lantern. If the dogs - a greyhound type and an Alsatian type - could corner the fox he would try to shoot it. Usually they would return with a few feathers or a chicken carcass but on occasions a second fox would raid the roosts while they were away and make away with another bird unhindered. Dad would usually claim he shot at something but the commotion added terror to a dark night. Once a dog was badly ripped on the shoulder by a cornered badger. Mother used to come into our room and comfort us at these times.

Dad's favourite breed of pig was the Wessex Saddleback. A rather docile lop eared breed which would eat grass and some weeds ... and even eggs if they came across a nest. His idea was to breed 'blue' piglets for sale as weaners at about eight weeks old. To this end he had about six sows and a white boar ... either a Large White with a sharp nose or Middle White with a snub nose. Sometimes the boar would run with the sows in a field or indeed several fields. But usually a boar was kept inside a building for safety.

Sows , like hens had a tendency to make nests in quiet secluded places and would, having been placed in a sty with adequate litter break their way out and make for the woods ... to have their litters. The tiny baby piglets were also a favourite item of foxy diet. They had to be found and collected before dark and Dad placed them in a 'willy basket' and they were taken with the sow back to a pen . An operation not highly approved by a sow at this stressful time ... and sows can give a hard bite. I would be allowed to watch this from a safe distance and when the piglets were loaded into the horse drawn putt to go back to the buildings I would, from the age of about six, be expected to drive the sow to follow, using a big stick.

There were other operations on pigs that I was not allowed to watch. Large pigs at grass would not only eat the grass itself but also the roots and so would turn the field to arable ... ploughing by pigs! To prevent this metal rings were put in the soft outside of the pig's nose, with the aid of ringing pliers. ... a process that the pig

highly disliked and would squeal vigorously and buck against the rope that was holding it. The other, forbidden activity was the 'cutting' of the young boar pigs. The meat from male pigs has a ghastly flavour unless the testicles are removed at about four weeks of age. To do this Dad, and an assistant, used to shut the sow out of the pen while the pigs were 'cut'. The sow would then become frantic to protect her young with dire consequence to anyone in the near vicinity.
Thus Mother was expected to devise means of seeing that we twins and any visitors had their attention suitably diverted indoors, or in the far fields, at this time. Hoping that later when Dad sat down for a subsequent meal he would not reveal what exactly he had been doing in all its graphic detail. Although this might depress the appetites and save on groceries!

Dad never got around to keeping his own sheep but would let the grass to a neighbouring farmer. So we would have a flock of about twenty yearling sheep of the Dorset Down type, usually for a few months in the cooler months. Dad did not find much use for sheep on the moors and it appeared that he had particular difficulty in controlling these animals.
The sort of dogs he kept were more used to hunting cattle and keeping off foxes and other dogs. The two dogs did not quite recognise whether they were to eat the sheep, of just chase them away. He never quite persuaded them to bring the sheep to him and the simple job of getting the sheep near a building to eat some concentrates may mean his running for several miles and yelling himself hoarse in doing little more than exciting the dogs .. who alone enjoyed the exercise.

Whinnie, the shire-cross-thoroughbred young mare, was there from the beginning. Her initial skittishness and bad temper simmered a little over the years. Old Flower her Shire mother finally gave way to Smart, of similar breed and then the addition of Captain, a Shire gelding (neutered male) They lived at grass most of the year and had to be caught and brought in to the stable to be harnessed. The lower field where they 'roosted' grew some magnificent crops of mushrooms

It was the cattle that were the mainstay of the farm. The sale of milk. calves, some yearlings and finally the barreners -cows no longer breeding- gave us some income and, almost paid the rent.

The breed was the native shorthorn with roan bodies, whitish legs and the short horns curved inwards. There must have been a bit of stray breeds for some calves were black. It was a dual purpose breed supplying some milk and calves that turned into beef at about three years. The old cows usually finished up as beef also. I believe Dad used a neighbours bull or used his own young bull bought as a calf. Twice each day about nine cows had to be hand milked and the churns taken to the milk factory at Chaffcome- nearly four miles of winding roads off the hill. - a daily job for Whinnie and the trap.

This journey, in fine weather was a treat for me -Peter wasn't very keen - and once a week we went the extra mile to the outskirts of Chard to get the Daily Mail - about the only family contact with the outside world. For most of the farm helpers very rarely left the villages.

At the end of the byre -where the cows were tied up for milking and feeding- were some brushwood faggots with straw bedding on top . Here, usually, there were three baby calves tied with a short rope around the neck so that they could not lie in their own dung. These were fed on milk after milking - often surplus 'new' milk from a newly calved cow or even milk from a cow that had slight mastitis.

In a pen , or in a small field the 'young stock' were kept. At least this is where they should live but they usually had other ideas and this is where the dogs came in. My very earliest memories were of Dad and the dogs galloping around the fields and buildings in hot pursuit of young cattle. The dogs barking and baying at times almost like foxhounds, and Dad following up with his whistle and his frantic shouting to the dogs. The heifers seemingly enjoying the game.
In fact the commands to the dogs that Dad used soon became a legend to Mother and us twins taking cover indoors. Dad's first aim was to get the heifers stopped from fleeing out of sight. The dogs were expected to head them off . Dad's frantic shouts of 'skitti ..fur...um' (get up a-fore them) would echo over the farm. But the dogs were not always successful and once drove a bunch of young frisky heifers four miles down country lanes before they succeeded. Dad would blow his whistle to recall the dogs at a long distance. The commotion and even danger to small children standing in the way was almost like a herd of charging elephants.
It seemed that the dogs so enjoyed this chasing of heifers that they often failed to obey commands (It is a fact that 'working dogs' are known as 'skittibeasts' in my immediate family to this day ... seventy years on!) .

A great feature of my early years was the animals breaking out from buildings, fields and, indeed the farm itself. There were six miles of hedges on the farm and Dad put in a quarter of a mile of barbed wire fencing. The warm weather would bring out the warble fly and most cattle would run to avoid them 'stinging' their legs.

But despite the problems they gave I gradually began to gain an understanding of the creature mind. So by the time I started permanent work at the age of fourteen I began to regard the farm livestock as my friends.

Chapter 6
TRACK OUT

When we arrived over the fields to the Farm in May 1921 there was no visible way of getting out; no track ... not even a path. The Farmhouse, unoccupied for a year, was once a couple of worker's cottages 'nailed ' together. These workers were at the far end of the old Estate and probably walked everywhere. After it was converted to a farmhouse local people said that the buildings were mostly used for fat cattle. Some older stock retained by the high stone wall that now surrounded the rick yard. Some milk was used to make cheese in the 'outer kitchen' of the house. As Mother had sadly found out ... the stink!!.. .

Mother was quite used to living in remote areas in India and South America. But Dad, with his experience in the 'Levels' decided that a proper road was needed. Thus from my earliest days the thudding of an iron bar, hammering and the rattle of stones being loaded in the putt .. all just below the garden.... were an interesting feature of my life. It started my lifelong interest in rocks and stones.

Of course Dad had no money to build a road so all work was to be a spare time job.and materials had to be home quarried.. Dad was never happier than when he could 'get something for nothing'. and the road was to be no exception . Like his stiles, his pig styes and, in fact our tree houses, they were to cost nothing ... except sweat and a little ingenuity.

The easiest place to build a road was across two level fields to meet the steep Purtington Road beyond the outer Pig Field. Home Field was also too steep to go up with cart or wagon. So Dad decided to go diagonally a longer distance across two fields to meet the Purtington Road at Lodge Copse.

He first had to plough a strip of turf about seven feet wide across both fields and then shift some forty tons of top soil to the lower side. .The hedge, half way up, had to be levelled and 'the gap' made. Only then could some 300 putt loads of stone be dug out and spread. There were two patches of soft clay, with the consistancy of hard cheese, in the top section and this meant extra big stones. at these points.

He used the old farm flat roller to roughly level the first level of stones and this made a tremendous noise like the ringing of a thousand bells. Then one day when the steam roller was working along the Purtington road he managed to persuade the driver to call down for a 'firkin' of scrumpy ... and bring his roller with him . Mother told us he had a couple of rabbits and a few eggs as well.

All the houses and buildings in the area were buillt of local stone. There were large quarries at Purtington and on the hill above Chard that supplied the flints, with sandstone quorns, to built the local houses. Dad had said that flints were not good for road surfaces as they shredded the tyres. The council had a facing of Tarmac on their roads.

Across the middle of the farm was a rocky ridge .{below the 600ft contour) This marked the lower point of the chalk that typified the top of the farm (up to 750 feet asl) This band of rock was some fifty feet thick in some places. and varied in texture from a concrete like mixture of tiny round stones cemented into solid rock to hard flat slabs.
Then there was patches of very hard sandstone with occasional layers of flint.. At the lower end - down by the old orchard - the rocks were above pure green sand..

The quarry revealed that there were thin vertical shafts down into the rocks into which tools could fall. Dad said that he once found another vertical hole when he cleared a track down the hill that took ten putt loads of quarry debris to fill.

In places there were fossils. Fine specimens of scallop shells, ammonites, limpits, bi-valves and echinoderms (sea unchin) . There were places in Tip Hill where there were gaps in the rock bedding and layers of beautiful calcite crystals had formed.

When not being used Mother would take us out to the quarry and Peter and I would have little spades and buckets. Mother called all these fossils fish and she, and even Dad, agreed that this must have been the seaside. Tip hill must have been a cliff I thought but what about the sea unchins found at the top of the farm. It was a bit puzzling, especially after Mother told us about Noah in the Bible.

There was another source of stones for the road. The upper fields were based on chalk And it seemed that where there was chalk there were millions of roundish flints. In fact the soil above the chalk is now known as 'Clay and Flints'. So stone picking was an alternative job to quarrying if anyone had spare time.

To my infant mind all this seemed very interesting and I started a lifelong interest in the Natural sciences. Here Geology, evolution and ecology form the basis of Natural History. I gradually got to know how the fauna and flora were distributed over the farm and surrounding fields and woodland.

In my early years the farm was teeming with wild life. There were areas of special interst to me. At first Mother or a mature guest would accompany us but then Perer did not share my love of nature. He was a great reader from an early age and showed an interest in electricity and mechanics.

I felt quite safe if I kept on the farm and the farm hedges on three sides bordered the public roads. The fourth side was the border with Lower Purtington Farm untill we got to the Ram; here started the high canopy woodland of Poole Copse.

The furthest corner of the Top Field was hedged off by gorse. The hedge was overgrown and there were rabbit holes in the bank. Yellow Hammers and hedge sparrows sang sweetly from the tips of the branches. It was the old Parish boundary. This corner was the only place on the farm that grew bracken and the soil was loamy with a few small rounded flint stones. This was a triangular field and n the corner against Hilly X was an ancient great spreading beech tree. Its roots hugged the broad bank of the hedge. From my earliest days I used to rest on the sunny side, along the hedge were large juicy blackberries we called Luscious berries. There was one particular plant that always grew at the base of this beech. It was mint scented with tufts of purple flowers on each stalk. It was, I discovered later, called Foot Stool Calamint.

The rest of the top field formed the beginning of the valley . Some of the stones in this field were like large pebbles and the loam was mixed with some gravel. This was a good field to hear Peewits calling and nesting. Along the Winsham road foxgloves grew in profusion and a single willow tree seemed out of place. This was the higest point for miles around. There were magnificent views down into Devon and the sea at Seaton Gap and north to the Blackdown hills and out over Somerset to the Bristol Channel and, after a thunder storm, once to the Welsh coast.

Over the gate , by a lovely berried holly tree was Higher Six Acres. A steep field noted for its mushrooms in the autumn. This field was grass and amonst its flowers were wild pansy and fumitory and red poppies. Adjoining 20 acres Copse - an ancient woodland with rare plants like Herb paris, twayblade and butterfly orchids - there was a steep fifteen foot bank down into Lower Six Acres. This bank tapered out along the hedge where a very knarled Oak guarded over the exit gate. The whole hedge was speckled white with rabbit holes ... for they were digging in pure chalk. At its steepest there were mounds of chalk and dried grass where a colony of badgers had tunnels up into Higher Six acres and along into the wood where there was a large covering of wood garlic with its starry white flowers.

I knew lower six acres for it adjoined Fairyland and was often used by Roe Deer and Foxes as they made for the great wood. A large rookery in the wood meant that these birds regularly over flew the field. In the centre of this field the soil was thin overlying the sandstone rocks. The growth here was sprinkled with quaking grass, yellow rattle and white clovers. The strange broomrape also grew on the margins with some very fine Spear Plume (Bull) Thistles. There were several species of Trefoils and the blue butterflies were always a delight. There was a sunken track along its north side.

Looking down on the two Six Acres was Hilly X . Tilted into the sun its Main Road hedge was dotted with Sycamore and Fir trees. It was said that before Purtington had a road out this was the horse track used to escape to the main road. Along the top was a smooth sided pit which was said to have provided chalk for spreading on the land and also providing material for the Lime Kiln in the Wood over the main road.

The largest chalk pit on the farm was next door in Whitwall as its name indicated . This field had a thin soil over pure chalk. Its hedges were solidly bedecked with primroses and it also grew some fine thistles of several species -Bull, creeping and scots. Visited by the larger species of butterfly this field produced wonderful flower meadow hay. The top hedge along the main road shared a long thin copse -Hebers Copses with Hilly X. It had some great silver firs beech and sycamore. I knew this hedge for the songs of the Woodlark , the Tree and Meadow Pippits and the everpresent skylark.. Doormice used the whole of this area.

The next field along the main road was Eleven Acres. Trees - holly, fir, sycamore and a great beech, lined its road hedge. I knew this hedge for its foxgloves and the occasional great mullein. This field was notable for the amount of sour dock and knapweed that it carried. At its southern side was Ten Acres which reached down to the entry of Fairyland with the tiny Bee Orchids growing in profusion.

Home field had two chalk pits, One a large affair with several huge hornbeam trees with their curious bird nest 'look-a-likes' in the outer branches. Patches of cowslips grew along the sides and several holly bushes carried masses of berries.

The next field was Hindhouse. Dad's road now divided this field as it did the final field Ten Acre Lodge, both fields had pits - for either chalk or flints in the corner. These fields had their woodlarks with their liquid songs. In Lodge copse the shimmering Wood warbler could be heard .

Below Lodge Field was the Outer Pig Field which lapped the newer Purtington road down to Pool Copse and the Cricket Estate ponds and streams. A worked out building stone quarry bordered the road and parts of all these bordering fields were steep and full of unusual plants and grasses and sedges. It was a great site for butterflies, dragonflies and strange insects

Further back than Poole copse two field skirted the valley on the other side. Navvy Show and Show Mead were steep gravelly soils with Earth Nuts and the Callous Rooted Dropwort on the steeper parts.

Chapter 7
FAIRYLAND
... Mother's Jungle

From the beginning while I was inside our rambling farmhouse I felt a degree of safety ... despite the open fires, the ill-fitting doors and the use of candles. Mother seemed always within call . But once outside we were in Dad's world ... unpredictable and full of dire secrets.

The dogs and other animals were so often tearing about at speed with Dad shouting and using his whistle to try to control matters. Then the poultry were often far from friendly ; with the odd cockerel sidling up to us with ruffled neck feathers or the geese approaching us with necks extended and giving a frightening hissing noise. A brooding hen with chicken would resent our attention to her offspring Every creature seemed larger and snarlier than we were. Even some plants .- such as stinging nettles and holly- would 'bite'.

Often on a fine afternoon Mother would gather us twins together with Judy, our spaniel and - like a hen with chicken - usher us outside. Mother carried her usual wicker picnic basket and we all carried our sticks ... selected as to size. We dodged any cattle in the yard and headed out to the comparative safety of 'Homefield'; (Dad called it 'Omefield' which made mother wince ... especially as he also called next door field 'Indhouse' to indicate its position behind the house.)

Across the lower end of Homefield were a line of pollarded ash trees. They looked very old and wizened with huge mightily wrinkled girths. Their roots spread out around the base forming a circular sloping bench. The roots spread further out again as if reluctant to force down into the flinty chalk soil beneath. Some of these roots were hollow and there were vertical chinks into the tree centres. Mother would look fearfully as she told us about poisonous snakes and scorpions in the tropics.

Don't stop at No 3!

Shared Breakfast

Secret Look-out

Is anyone in there?

Landing Gently by **Paul Smith**

DOWN THE LANE

HOUSE IN THE WOODS

OUT OVER SOMERSET

ROAD WORKS

Landing Gently — by Paul Smith

Each tree in its crown sprouted ash poles from its leafy summit. From here .brambles and ferns scrambled with even the odd seedling of elder and sycamore peeping out - more like Bonsai trees - from a precarious footing.

Small children and small dogs were magnets to most cattle but if the yard animals were not loose in the field it was here that Mother would rest a while on the natural seating while she would tell of the dangers of sitting on trees in India.. We were only interested in seeing if we could find strange beasties as we poked away with out little sticks.

The five trees were spaced in a straight line across the field. The first tree had a large gash into the hollow centre and we sometimes managed to extricate a stag beetle or a centipede with our sticks to Mother's cringing horror. The second tree was undamaged about its bole but one root was hollow back into the tree so that rabbits could enter and foxes and badgers would leave their claw marks. We would listen for inner noises,

The third tree had a massive top with many hollow roots into the base. Mother always made us avoid this one. The gruesome mystery took many years to emerge. The farm cats used to produce several litters of kittens during the warmer months. But these nearly always disappeared before we could make friends with them . Then one evening I saw Dad throwing small objects into the greenery of this tree. The fourth tree was a mystery to us at first for strange mewing noises would emanate from its centre at around dusk and dawn in the springtime. But Mother would not let us touch this tree as she said it might be bats as well. It turned out to be little owls.

The fifth tree was slightly shorter than the rest and, from standing further up the field it was possible to see that ferns completely covered the top. If this tree was tapped with a stick in the spring a stock dove would noisily flap out from its nest in the summit.

Mother, at first, tended to dress us in white clothes so that any mud we met was readily visible. Near the buildings even a slight storm would turn the chalky clay soil to mud which would often become mixed with animal droppings. Mother's solution was to take us further away. Through the Homefield gate into 'Ten Acres' and along a rough stony track to the far hedge and then down along the hedge to a stile - one of Dad's special no-cost rustic models - and into a long straggling coppice of great variety we quickly got to know as 'Fairyland'.

Beyond the stile was a small ridged grassy bank where Mother would sit and tell us of her foreign adventures or read to us from fantasy books. This was our base and she would stay there while we went down one of several tunnel-like paths into the shady depths of the wood. Nut bushes 'caressed' the floor and the paths meandered around bushes and patches of brambles. The high canopy of oak, fir and ash reached, to my small mind, right up into the sky. The floor had several earthy banks - marking old hedges - and there was a sunken road across the middle for us to explore.

In one part there was a rocky outcrop with a steep bank with tiny plants with bright flowers clinging to cracks between the stones. In several places throughout the wood the sun could reach the floor to nurture primrose banks with foxgloves - here were the butterflies, bees and other busy insects. Fairyland was surrounded by fences or hedges and we could peer our on to fields from this living 'hide'.

If we were quiet or split up we could sometimes see deer or foxes in the fields. From here we could also observe our own cattle or horses grazing or resting. We always within 'crying' distance of Mother and Judy could be asked to go and find us.

The high canopy was the abode of a few Red squirrels. Crows and magpie nests were in the tops. In fact Fairyland was teeming with wild creatures. The floor was peppered with holes - from the large entrance dug out by badgers to the modest openings of rabbits down to the tiny entrances made by mice, voles and insects like bees and beetles. There were numerous places to sit and watch and listen to the natural world around us.

Because some trees had rotten branches on the floor and some trees had been felled and the trunks dragged out by horses, there had grown up many different fungi. There were many delightful sorts of toadstool and bracket fungus - including the strange growths (Jew's Ear) on the dead elder branches . Mother told us not to damage any of these as elves and fairies would use them for seats Many of the stories Mother told us enabled us to build up a mythology about the wood ...and that is why we called it Fairyland. Mother always carried her large basket containing some drinks and a blanket ... and even an umbrella. On the return journey we would all be loaded with goodies as to season.

The autumn - the time of bounteous plenty - would see us with nuts both the elegant 'filbert' and some thinner sorts. There were several grades of blackberry - from the small harder variety to the 'luscious' berry. Mushrooms - the common sort and occasionally the giant horse mushroom from a neighbour's field adjoining the woodland. With the rocky outcrop and the old cart track we would pick small fragments of rusty metal and a range of small flint and sandstones with patches of blue, red and white.

We often managed to pick Mother a small bunch of flowers. Perhaps primroses . bluebells to which was added the occasional rarity such as the bee orchid or twayblade. We always had to bring back some firewood sticks. ... from the tiny lighting twigs, chippings of larch we called 'firelighters', , to larger branches that I - and Peter - would have to drag back to the yard to be chopped up.

As I got older I was able to enter Fairyland by myself and climb up some of the trees, to go bird's nesting to find the nests of new species . I would only take one egg from a full clutch of a species I had not already got in my collection.. After many years the cattle were able to use the woodland for shelter and much of the old magic had gone.

It was indeed the magic of Fairyland that started my lifelong interest in the wild creatures around us and the wonderful woodland sites that are their home.

Chapter 8
THE RAM

Like all tiny children my early life was full of mysterious happenings. Mother was good at explaining why most activities took place under her control inside the house. But there was one little ceremony that Dad did in the house that puzzled me greatly. Each day - morning and evening - Mother would tell us to be absolutely quiet while she held her finger to her lips. Dad would go to the kitchen and - underneath the roof water tank, place his ear to the rising pipe.

After a moment it was over and he would usually go out on the farm ... but occasionally he would grab his cap and rush down tip hill and out of sight.. All Mother would tell us was that this was the Ram. I listened to the pipe myself and could hear the regular faint thumping, as if an animal's heart was beating .. it was a sort of double beat. The pipe was always cold to the touch and in warm weather beads of water dripped down the wall nearby.

Tip Hill was a forbidden place to us twins, and thus this mystery gained more significance. Certainly Mother did not volunteer an answer. Perhaps because the finer points of water supply in far off places were certainly far outside her remit as a nursery governess. Dad never allowed anyone, other than himself, to tend this mysterious beast called the Ram. .

The countryside was always full of noises to my sensitive ears. Cattle, pigs, poultry and other attendant livestock vied with the rustle of the wind through the foliage. But from the furthest point of the garden it was possible to hear a faint double beat of the Ram. It obviously lived in the watery end point of Pool Copse. Sometimes in the comparative quietness of the night this double-beat would also vie with the squeak of bats, the calls of foxes and other dark denizens of the hours of darkness. During our earliest years Mother had naturally tried to keep us together but my more primitive interests slowly began to emerge and it was the day that

Mother allowed me to accompany Dad down Tip Hill to the Ram - at the age of about four- that broke the spell.

It was about mid-day with the summer sun shining over the luscious countryside when Dad collected me from the front door and I, and Ben - his lurcher dog - went through the lower garden gate to Tip Hill. .The path was steep and winding with rocky outcrops but we both had sticks to steady ourselves. Dad also carried what looked like a thick rubber ring,... perhaps as 'food' for the beast!

At the bottom of the steep hill we entered the old orchard dotted with gnarled and twisted apple trees and a few high growing marsh thistles. In the corner was a pad of concrete with a chain emerging from its centre and the end tied to an iron ring. The full sound of the Ram's song was now audible and Dad told me it repeated 'Pull.. Up .. Pull .. Up ... Pull.. Up' continuously as it drove the water to our farmhouse at the top of the hill.

Dad untied the chain and immediately the 'beast' became silent. He said I was to follow Ben who was 'at heel' as he climbed over a stile - Ben and I managed to squeeze underneath the lowest bar - and there was a little path that led into the shady end of Pool Copse. There was another concrete pad with a heavy wooden trap door with a padlock . I sensed that this was the cage of the Ram ... I was already feeling sorry for it. Dad opened the trap-door and let it fall back on the concrete with a clang. He lowered himself slowly into the chamber below and I could hear a splash as his feet touched the water. As he disappeared into the chamber he told Ben to 'stay' and so Ben and I stared into the darkened interior. The water on the floor was rippling as Dad's feet moved around, casting erie moving patterns on the smooth walls. I was glad to have Ben with me as my eyes quickly got used to the darkness of the chamber.

At last I could see the Ram. It was a dark metallic green with a rounded body like an octopus. Silvery pipes, like tentacles, spread around the chamber. Its mouth was a sort of circular grid and I could see that Dad was fiddling with the rubber ring that he had brought down with him . After only a few minutes Dad moved us back so that he could lever himself out again with his hands gripping the sides.

He told us to 'stay' once more as he made his way back to the chain. Suddenly there was a gush of water from the 'beast' and it hesitated for a second before the familiar contented 'pull ...up' 'pull ...up' 'pull .. up' again. echoed out over the woodland scene. At each beat a spurt of water gushed out only to suddenly stop and the process repeated.
From the Ram chamber an underground pipe took the water along the bank to join the stream coming from the direction of the wet marshy area I knew as the weir.

After Dad had cleared the outlet of a few small stones we all went back out through the old orchard to another small stile along another bank to a low concrete wall that was obviously holding back a small pond. Ben and I balanced on the flat top of the wall while Dad pulled up a wire grid covering the end of a pipe under the water.

Watercress was growing in the stream here and Dad gave me a handful to hold on our way back for Dad never returned empty handed and usually found a fallen branch of a tree to carry up Tip Hill.

This was my first trip to see the Ram and it was not many years on when I learnt that this was a very exciting area. The stream started in a pot-hole just inside the top hedge when in times of a flash flood water rushed down the valley beyond and entered a hole in the land with a six foot drop. This pot-hole was never dry and was full of tadpoles in the early summer.. Down the copse - the weir- other little wet area seeped in more water into the main stream. as it meandered through the dense undergrowth.

There was a low slot in the weir wall to take an overflow down the stream and past four tall aspen trees that stood in line with their branches arranged in whole up the stem. as Mother never came down this way I could escape and climb some sixty feet into the summit with comparative safety.

I gradually knew every inch of the Ram area and often climbed and sat reading in the trees. I always had the 'song' of my friend the Ram to keep me company.

Chapter 9
AROUND THE COUNTRY LANES

It was perhaps a good thing that we had no near neighbours ... so no tongues to start wagging. But every possible Sunday afternoon ....dinner completed ... Mother could be seen fleeing the house and making hastily for the nearest public road, with us twins - and sometimes Judy - in tow.

At first we were in her increasingly battered double pram but later we were on foot or our tiny Fairy Cycles. Leaving the house we had to negotiate the very rough stones of the farm track , alternatively two muddy gateways and across two fields, to the Purtington road. This itself was surfaced with gravel rolled into tar on a stone base.... with a top dusting of finer gravel. Not very easy on small wheels..

This walk was the one domestic concession Mother accepted from Dad. He insisted that he had one hour a week when he could sleep. without disturbance. He would sit in his armchair by the fireplace with his feet up against the mantelpiece for this purpose and in order to sleep in this uncomfortable position he required complete silence from us twins. He reckoned that we twins were incapable of being silent for that period. He was probably quite right for if we were not chattering to one another or talking to Mother we were talking to the animals or just .noisily playing. To a small child if something is not making a noise of some sort it is not worth playing with. Dad did not seem to mind the farming and wildlife noises that continually rent the air ... from birds and insects to cattle mooing. .

So Mother- with blinding logic, decided that we desert the farm and explore the many narrow high banked lanes that tumble off the main road that crosses Windwhistle Hill. This would also take in our neighbouring settlements of Chillington and Purtington. and more rarely Wayford. Mother would not herself dress for walking but had a long dress buttoned to the neck. We wore short thin trouser suits.

The two-mile Windwhistle hill ridge twix Chard and Crewkerne carries the old London to Penzance main road east/west at 700feet above sea level with stunning views over much of the rest of Somerset and across the hills of Dorset. Leafy narrow lanes plunge to the north down through scattered houses for over two hundred feet before small streams emerge from the clay at the base of the Greensand.

To the south there is much chalk and the lanes undulate though several valleys before reaching the River Axe three miles distant. Here again water appears at some 500 feet.

The Purtington road had gates at every field and there was the last gate, between two small stone walls, by our 'ten acre lodge' field. Mother would normally try to reach the main road for us to perhaps see a few cars - an Austin Seven, a Morgan or a Ford Model T. There were sometimes. lorries and the London Royal Blue coach would pass on weekdays. Horse drawn vehicles would also use the road. We could hear vehicles coming from a mile off so as the road was just wide enough for two vehicles to pass Mother would get us on the verge so that we could wave to the drivers.

Our greatest treat was when the road was being repaired. Here steam traction engines would pull four wheel trailers with stones and gravel. One pulled a tar cylinder with a fire underneath and the tar was laid hot with the stones or gravel laid on top. A steam roller - our great favourite - would slowly finish the job before a light dressing on top. Amongst all this steam power and the glorious aroma of hot oil, steam and tar was an large elderly man - Mother said he looked like Mr Punch - riding on a small cart drawn very tardily by a sad looking greyish mule. He apparently hauled the fine gavel to finish the job. .Although the Roller driver smiled and waved to us Mother was a bit wary after she heard a bout of bad language coming from the other men .. who seemed to become more spattered with tar as the day wore on.

The eastern mile of main road adjoined our higher fields and towards the end was a thin strip of copse with high trees of beech, silver fir above a dense floor covered with hazel, elder, privet and wayfaring trees. A badger and deer path went trough the centre over a bed of wood garlic or bluebells. Mother let us use this path -if we did not pick the garlic - as she stayed on the road. This was called Hebers Woods.

Opposite was the very steep Heber's Lane. The banks were high with high trees and half way down was a small wooded patch and a couple of tiny houses - Mother said an old man was said to live there ...but we never saw him, He would have had to go a mile to get water from a stream. Down further was a cross roads (crosslanes) . We took the steep west lane with steep partly bare banks and a variety of wild flowers and toadstools. The whole area around was of small fields, tiny woodlands and great beech and oak trees skirting the lane. Mother never hurried us and we could sit and absorb the sheer beauty and isolation of the scene.

As the lane began to rise again we could see the only water supply for the few houses before us. It was coming from a pipe poking our of a steep laneside bank. A rough concrete backing had been made and a small standing for buckets to stand to catch the water. This was higher Chillington and a carpenter and wheelwright were busy in a workshop one side of the lane. Mother would let us peer in to see them making wagon wheels or repairing wagons or carts At the far end of his works the unmistakable screw thread of the local cider press was .awaiting the autumn apple crops. Opposite was the blacksmith with his smoky inferno and a pile of metal rods resting beside his door. If he was not hammering metal for fitting to farm machinery he was also the local farrier and the acrid smell of marking horses hooves with the hot shoes would permeate the air. Both men - Mr Bussel and Mr Hodder - had young children of their own and would stop work momentarily to raise their caps with a "Morning Missus ... Aaah! they twins be growing!"

Further up the hill a group of three basic cottages hid behind a high hedge and were approached by a track - these were known as 'Four Gun Battery'. But no one knew exactly why? There were a few other houses further down a steep side road .. one of these was the local Cobbler - who I gradually learnt - could work wonders on worn out shoes and could often be seen , working at his last inside his open door. The lady of one house down this lane had converted one front room to a tiny shop selling some items for survival in these parts..... such as candles, soap, brushes, fly-papers, mousetraps and boot polish as well as locally produced foodstuffs.

As we slowly went up the hill towards the Main Road we could see a few more simple tiny local stone houses peeping out from the abundant trees and bushes. Mother told us that the one in the wood belonged to the village poacher.
Where the Chillington road reached the Main Road there was the huge pile of flint stones with the stone-cracker (but not on Sundays) and a house just further on. The latter held a notice that SEYMOURS fizzy drinks were on sale.

So back to the Farm, with at least some booty such as blackberries, sloes and always a bit of firewood, where Dad was by now doing the afternoon milking . We were weary and it was our turn to sleep.

Of course Mother often went to the shop, the cobbler or even the carpenter with us during the week. Then we often took a short cut straight up across the steep home field and across the main road into a piece of wild public unfenced field called The

Downs. This was a place where a couple of gypsy caravans were sometimes parked and apart from barking dogs the sight of twins seemed to attract their womenfolk ... perhaps they were sensible enough not to rear twins, I thought.!

As we grew older -from about six onwards- we had our fairy cycles, and at weekends and in the school holidays Mother was more ambitious, we could then do the five mile trip to Lower Chillington which had another very small shop and had a tiny ancient Church and even ran a school.. Pushing the bikes up hill meant less than half was on foot. Mother's bike also had a carrier and a front basket. She must have been strong for she would even take one of us the nine mile round trip to Chard.

Another trip all three of us did was to see a very old lady in Wayford - about a six mile round trip with one long hill. here we had a cup of tea or lemonade to refresh us for the return journey. We continued to do the shorter trips on Sunday afternoons, in a sequence but I was never bored.

It was when Mother got her first Austin Seven and we attended school in Chard that our isolation and our trips around the local lanes came to an end. I was by then quite independent and would go alone for long walks on the farm . . .

Chapter 10
EDUCATION

Shortly after the turn of the century (1900) Mother was completing her nurse's training at Bristol Infirmary and Great Ormond Street. These were the great days of the British Empire and she, and her sister Lilla, had no difficulty in signing good two-year contracts to act as Nursery Governesses to the very rich . She would relate to us tiny tots the adventures of her tours around the World. India, Argentina, Denmark, France. The sumptuous households in which she worked consisted of cooks ,maids. butlers. grooms, and gardeners. There also were the exciting excursions - in horse drawn carriages - into the local towns and countryside There was usually only one child of the household. In that society it was probably thought indecent to have more. The infant had a wet and dry nurse and then at some two years of age a Governess. The child was highly pampered with medical help always on hand. There was a special nursery suite. Mother lived with the child and was expected, by the time an official boarding school was arranged at about six years, to have produced a perfectly mannered, correctly spoken, well behaved, utterly clean and tidy .child. With , of course, all the snobbery and social understanding so valued at that time. The child would normally be taken once each day to see its parents and be inspected and questioned. If the family's language was not English then the Governess would be expected to also teach the parents.

Having completed a long spell in India Mother joined a diplomatic French family for the Great War. At the age of forty, returning to her family home in Bridgwater she decided it was time to marry and have her own child. But Nature must have smiled at such impertinence ... the result Twins. Dad was a small holder in the Village.

So my education thus started at the time when we left the watery wilderness of the Somerset 'Levels' and set out on that sunny May afternoon in 1921 for Windwhistle hill and the farm. In those early years Mother tried to replicate in us those desirable qualities of her former charges. We then were conditioned to recognise two classes of people. Mother and us twins .... and the 'natives' , which included Dad. This isolation suited Mother and she devoted herself with a selfless zeal to her task. Never ever voicing any regret. She had a small selection of children's books. which included Grimms Fairy Tales, most of Beatrix Potter, a coloured book of birds and the inevitable Bible. She did not rely on books for, as I well remember, she had soon transformed the farm into a place of mystery and magic with her tales of her life abroad. There were the wild and dangerous hills of India, the carriage rides in Bombay, the remote life of an Argentinian 'astansia' . Her War experiences as 'Big Bertha' pounded the outskirts of Paris while the family sheltered in a cellar. She had crossed the North Sea with Warship escort four times and told of her adventures in Denmark and Greece. We were never bored in her company and she soon created in our tiny minds an ever exciting, even if dangerous, international world of strange exotic people and wild beasts.

There being no outside influences other than a few bits of 'news' from the headphone of a 'cats whisker' wireless and one newspaper - usually the Daily Mail - each week; we soon learnt to read , write, draw and paint.

The schooling facilities around the sparsely populated area of Windwhistle were not promising. There was a tiny school at lower Chillington at some two miles across fields, the "downs' and steep roads. Four miles to the south was the Village of Winsham. Here was a school catering for all ages.

Dad had attended such a village school until he was eleven - when he had to take over a small holding to be breadwinner for his mother and two younger sisters - Mother seems to have good reason not to 'inflict' this type of schooling on her so carefully nurtured twins. Dad believed that school had never done him much good and was now too busy to take any interest. Apparently Mother's professional charges had either a private tutor when she left them or were sent to a boarding school that catered for wealthy parents. Luckily for us twins a solution was at hand. A larger farmer, who owned the hamlet of Purtington - had five children . The wife, Mother's friend Elsie', employed a girl teacher who lived in the servants quarters of their house on the hill - a mile distant. There were only three children at home - two were at boarding school - so for the next four years we progressed under the gentle tuition of 'Miss Jones'.

Dad was too busy to provide transport ... a horse and trap... so we had to walk. This mile was full of interest, a sort of ever changing natural history lesson in itself. We would start off through the garden gate skirting Tip Hill. Here we could overlook the 'waterworks' and the old orchard with its ancient twisted trees and its pithy cider apples. Across a style to the 'pig field'- so called because it was the only field subjected to rooting pigs ... that was when their nose rings had come out.

At the lower far corner were the 'rabbit holes' - a mountain of fine greenish sand dug out by a colony of rabbits , badgers and foxes from time immemorial.
The path then became a bit tricky. Another stile in Dad's own rustic design and a steep bank - known to have the first primroses of spring - leading through a hole in the hazel hedge leading onto a public road. From here we could see the great house peeping out of the trees on the hill across the valley.

But first there was a long roadside wall , full of tiny colourful flowers and ferns, that skirted Pool Pond. There was a point along this wall where I could pull myself up to see the water just below the top. There were always moorhens, ducks and often a kingfisher. There was a large marsh on the other side below the road. Great stalks of mare's tail vied with water dropwort, marsh orchids and great willowherb.

There were some huge trees towards the far end of Pool Copse. One sweet chestnut spilled out its ripened nuts onto the road below. A little further along a farm track led along to a gate that was often closed. Further along as it looped to go up the hill, it enclosed a mill pond with a race, to work a water wheel working a circular saw used to cut up firewood and make posts. Beyond this the stream had several concrete based pens used for sheep dipping.

Mother allowed us half an hour for the journey ... with strict instructions not to leave the prescribed path. We used to arrive at the back door and pull the bell so that Miss Jones - we had to address her as Miss - could see we got ready to enter the classroom.

She appeared to me as a very young version of Mother. She was bespectacled, impeccably mannered and spoken. She was trained to teach children up to the age of eight and we were told that she was then going to enter a nunnery. She had obviously been carefully selected as the family were Catholics.

We boys were ourselves now reliving the sort of life Mother had put upon her foreign charges. We also had a daily visit to the schoolroom from Elsie whose entrance was as purposeful as it was formal. The father used regularly to appear to announce he would be out till dark on the farm or market. If Elsie was there he would use his Irish blarney to say how lovely Miss Jones looked and thus receive a scowl from his good lady!

Miss Jones could give us individual attention. It was mostly reading, copybook writing and a little arithmetic. What delighted me was a period of art. She got us to draw and paint flowers or a more formal object like a teapot or tea cup and saucer. She would help us to paint a bowl of fruit and put the reflections on a ripe apple.

As the other children reached boarding school age they left until only us twins and the youngest boy remained. Mother had to do some deep thinking. We were now coming up to eight years old and by now there were several means of 'escape' from the farm ... other than by foot or horse. Mother - with financial help from her now married sister - had her own kingfisher blue Austin Seven which resided in a spare pig's sty. Dad had a Ford flat bed lorry with hard wooden spokes and solid rubber tyred wheels. And we boys both had bicycles. .

That summer saw the arrival of Aunt Jo ... complete with a stray 'mother and child' to see to her every need. We twins were, as usual, paraded and questioned to see that we had not slipped in behaviour or cleanliness. We heard her discussing our future education with Mother.
Aunt Jo stayed an extra four days that year and we had another picnic -complete with disaster. This time she sat on an ants nest and we all got bitten. That necessitated some very undignified scratching in unmentionable places!

She had indeed prolonged her stay so as to join Mother and us to visit the Headmaster of the boys 'public' school in Chard. Chard School was a boarding school where rather wealthy Service officers , or people living abroad conveniently parked their children during term time. .

The School , presumably to fill class vacancies , did 'tolerate ' a few dayboys. As a school it was very strict on manners, had a ghastly 'fagging system' and believed in the dictum that vigorous activity on the sports field left boys little energy to breach rules.

The Head - a large red-faced middle aged man who lived with his mother - dealt with any wrongdoer at mid-day on each Wednesday when he would exercise his array of canes on their bottoms as the victim bent over a heavy oak desk in the lower fifth class room. The crack of the cane could be heard throughout the school.

Peter and I had first to serve a probationary two terms in a junior section across the road from the main School. Here - and only here - was to be found a woman teacher. As well as learning the basic rules we had to assimilate the custom of never using Christian names. As there was another Smith in the school - Smith major - we were Smith minor and I was (by five minutes) Smith minimus. .

From the beginning our transport was a problem. Mother arranged for us to forgo sport after school .. to the horror of the Head. So we variously arrived at School at 9 o'clock each day by Austin Seven , bicycles or Dad's ancient and grubby lorry. In the latter case usually late and noisily deposited outside the Head's study. The lorry also carrying a couple of milk churns, slabs of cotton cake and a few sundry hessian bags and machinery parts .. even pigs or calves. I believed the Head quickly regretted taking us on.

My six years as a pupil at Chard School was for me as hard as it was puzzling. I was always given the impression that I was not really wanted .. a sort of outside leper who crawled into the classrooms when no one was looking. The boarding pupils were forbidden to leave the School grounds ... if they should so much as speak to a town girl on the outside it merited a severe caning from the Head.

The boarders were given two hours of homework every working evening after games in well lighted classrooms. We had to manage with a very temperamental Aladdin lamp or the flickering light of a candle. I had the lingering feeling that to skip games and not doing proper homework would lead to a caning.
It was in my last year that one of the older borders suffered a flogging in front of the entire school .. including the Matron and the junior mistress .. followed by expulsion. He was said to have pinched some money from a visiting cricket team member.

Each year the School put on what they called a 'Gaudy'. A sort of open day for parents to parade themselves' their cars .. a Rolls Royce if possible . and wear the finery of clothes and jewellery that befitted their station in life.

While Dad stayed back on the farm Mother , who knew quite a bit about such social activities, parked her Austin well out of sight and accompanied us until our extra pocket money ran out. She was not formally introduced to other parents.

During these school years I got to know quite a lot about the strange mixture of my class mates. Of the fifteen, three at least claimed to have millionaire parents and they hailed from Canada, Brazil, Guatemala and another from Afghanistan. Several had parents in the forces. There was one strange boy who never had a home to go to for the 'hols'. He hinted that he was the result of an 'indiscretion, by his wealthy father. Peter and I were always kept in different classes.

There was another factor that divided us from the vast bulk of the pupils. Mother had, as soon as we got to the farm become Roman Catholic ... under the influence of her sister.

So it was arranged that -during the one weekly period of religious instruction - we should accompany ,usually, two boarders across the road to the priest who lived in the Presbytery down the road. Although I was happy with this arrangement there was a problem for boarders. There was a convent school attached to the Catholic Church and the pupils here were doubly contaminated .. they were both girls and Catholics.

Of course the Priests who instructed us - mostly in the catechism - did not share this view but the St Gildas heavily veiled teaching nuns had to shield their charges from a possible male pollution. They must have thought that fewer pupils would subsequently seek to be nuns themselves.

The teaching at Chard School was aimed at passing the Cambridge School certificate at sixteen years of age. All our classwork was based on set books which were unbelievably dry and uninteresting. With small classes we had regular written tests and had to read to the class from the books. The atmosphere was unsmiling with little opportunity for self expression. For art we drew cubes, cones and rectangular blocks on a white surface. It was quite a change to draw even a teapot. Failure to show the required diligence in class could lead to up to a hundred 'lines'. More than two hundred lines in any one week would also merit a visit to the Head for caning.

For our first few years we had dinner with all the school in the refectory. This was a highly disciplined affair almost like a military parade. The grace was said in Latin - a compulsory subject- by the classics master. In his absence another master would use the short words "Benidictus benedicat .. Amen"

But as the great economic farming recession of the 'thirties' took hold we took sandwiches with us and got a hot drink at a Chard cafe. Dad could see no reason for us paying school fees but Mother took on extra paying guests and Aunt Jo may have also helped.

As I approached my fourteenth birthday I had to take a vital decision . Dad could no longer afford to pay the wages and the last man would have to leave. The farm would have to be sold. This could be avoided if I left school immediately and worked full time on the farm for just my keep ... no wages. Like so many of the unemployed farm workers I would be free to earn some pocket money from the abundant wildlife around. I could not yet 'have the rabbits' as a skilled man caught our rabbits on a fifty/fifty basis. .

Although Mother had until now resisted Dad's desire for us to do real farm work I fancied myself as a labourer and a had even marshalled a few of Mother's guests

Landing Gently   by Paul Smith

NOT ALL ARE TOO FRIENDLY

OFF WITH THE MILK.

AWAITING THE MILK LORRY

HAND MILKING

ANOTHER WORLD

POOL COPSE

WHAT! NO TROUSERS?

IN THE CANOPY

NIGHT SHIFT

THE HURDLE MAKER.

LANDING GENTLY   by PAUL SMITH

to help stooking corn, collecting firewood or gathering nuts. I could even call recalcitrant animals 'little buggers' like real workers.

In fact my decision was clinched when Aunt Jo's solicitor husband sent word that he would employ me as an office boy and I could live with them. Despite a warm lavatory seat the idea of living in London made me shudder with horror.

So it was that at the tender age of fourteen I passed my school uniform to Peter, who had no interest in countryside things, and donned an old cloth cap, some old long trousers that mother had given to her. If rain came I would drape an old sack around my shoulders. I probably resembled some waif out of Dickens.

Mother had always accepted that I would join Dad on the farm at some later date So when a Young Farmers Club was to be formed in Chard she had driven me to an inaugural meeting a year earlier. I even now had a calf, supplied by the Club, to rear in the approved style and I had invested some pocket money in a mole trap and a row of skins were now drying on boards.

It was a shock for me to rise at 6 o'clock every morning to get the milking done. This was a painful process as the cows at first did not appreciate a small boy sitting on a three-legged stool and trying with small weak hands vainly to squeeze, often in the wrong sequence. Dad had been instructed by Mother to see that I did not call the cows 'them buggers' even if I was kicked or even manured upon.

I soon learnt that practical farming was lacking in glamour and was very hard on my child's bodily frame. Worst of all was helping to hold piglets for castration, ringing adult pigs and calving cows. I knew that animals had to be mated before they produced young but did not quite realise how dangerous it could be with a rampant bull or boar let loose.

By 1933 Dad's old lorry had become redundant. A MMB milk lorry collected the churns from a stand on the Purtington Lane. I had the tricky job of catching Whinnie from the fields each morning to be harnessed to haul out the milk churns in the putt.

My education in the art of farming was just starting. With the help of early farming magazines, lectures and farm visits with the Young Farmers Club I was always encouraged by Mother to better my abilities.

The story of Dad's farming will be related in another chapter Mother also encouraged me to take an active interest in other matters. This was accelerated after my old School tutor told Peter that after two years of work on a small farm I would become totally illiterate. This, at the time, annoyed me greatly.

I had from an early age been keen on drawing and painting. Particularly poster work and cartooning so I started a Commercial Art correspondence course with the Dominions Correspondence School. Much of my study was done in the evening by lamp or candlelight. I also determined to continue writing essays ... my favourite subject at School

Great was my delight when I won the national essay competition of the 'Young Farmer' Magazine a year later. They also started publishing my cartoons a year later. This later strangely led to a rift with Dad.

One day Mother had an unexpected visitor. He was a longish haired older man with thin sensitive fingers. He was the Principal of the Yeovil School of Art. He had seen my efforts and calling me in said I should attend their four-hour 'life' class on Wednesdays in term time. Mother agreed as I had learnt to drive the car.

That evening we told Dad of our idea. Without realising exactly what I was to draw he reluctantly agreed provided I still did my share of the farm work both before and after my journey.

So on Wednesdays, in term time, I got up extra early and first set out for my rabbit snares to bring in the night's catch. Then I did the hand milking, and calf feeding before a hurried breakfast, before taking out the milk churns to the stand and after some cleaning out and feeding penned animals.

A quick change of clothes and setting out for Yeovil. As I was the youngest student, some were professional artists, I studied in the skeleton room and then the plaster cast room for a term before, with the permission of the Principal, joining the others in the 'life' room at my allotted easel.
Here we drew in a temperature of ninety degrees - to avoid the model getting goose pimples - I and six others studied in detail the female form under the guidance of a tutor. It was a complete contrast to the art I did at school. Although Mother knew what I was drawing we kept it a secret from Dad .. whose attitude to women was very traditional and basic, and he would have much preferred I learn to hoe turnips.

After a couple of years farming was coming out of recession. Peter had left school and after a year studying wireless at home had left to live and work in London. By the beginning of 1939 it appeared quite clear that War was coming. I gave up my art studies and joined the Territorial Army field artillery in Taunton. They paid my car expenses and an hourly drill per week. They also gave me a camping 'holiday' on Dartmoor.

With the outbreak of War I was immediately called up and for me it was back to school. For the next seven years I spent much of my time at a bench or a desk. I thrived on frequent exams. The army no longer flogged soldiers who failed exams, they merely threatened to post them to a list to replace battle casualties.

Then it was for me back to farming.

Chapter 11
PAYING GUESTS

The old farmhouse was rambling and, to my infant mind, quite vast. Upstairs there were five rooms which Mother saw as potential bedrooms. They could be equipped with washstands - water jug, basin and soap dish - and the universal chamber pot. In fact she seemed to have collected a number of decorated pots on her travels for they depicted floral scenes of some beauty ... although she told us both that it would be rude to be seen gazing into a chamber pot ... especially as it might have been used.. Perhaps that was the privilege of the grown-ups!.
Despite the considerable difficulties and potential hazards that beset such a venture Mother started from an early stage to take in paying guests. They did, presumably expect to live 'as family' and shared the hazards of this isolated and primitive existence .. but Mother felt it only right to try to shield them from Dad's more extreme domestic habits.
Thus she tried to ensure that he went no further into the house than the 'outer kitchen' without changing his boots. Even then, as newspapers were scarce and valuable as fire lighters, sacks were put down in winter even in this room. A basin of water was also kept nearby for him to wash his hands at meal times.

As he did not always observe these niceties in times of stress, she took advantage of the fact that he was usually not on time for meals and so had to arrange for him to be 'refuelled' elsewhere. As he appeared to be either working or sleeping most of the time this also applied to the rather sumptuous evening 'dinner' that Mother would put on for occasions when several guests were present.
On one occasion Dad turned up a little late for such a special occasion and sat down with congealed blood down one side of his face. Not only that but he had used his special method of treatment. This consisted of applying a cobweb to the wound and letting the blood thicken on to it. On this occasion he had met a cows horn as he gave her an evening feed and had not time even to remove the rightful owners of the cobweb; one large spider was sitting on his jacket collar. Several guests turned green as Mother grabbed DAD ushering him to the kitchen tap.

None of Mother's guests ever seemed to be country people and so they held a rather idyllic view of farming. On another epic occasion Dad rushed in late for a communal meal and took a seat next to a rather sedate older lady.

"Aaah Lan," she cheerfully asked " and what have you been doing?"
" We've bin cutting pigs today", he replied.
"And how do you do that ? she enquired , perhaps not hearing him properly.

Dad had half explained to her the finer points of castrating pigs before Mother was able to distract his attention and turn the general conversation on to haymaking. The lady did not appear to eat the rest of her meal.

Of course more dramatic things did happen from time to time. Due chiefly to Dad's failure to close doors. On more than one occasion an entire litter of piglets entered during a meal and dropped their 'visiting cards' under the table. before being ushered out through the kitchen to meet their sow who had upset a sack of potatoes. At least animals - mostly calves and piglets - stayed underneath the tables. This was far from the case with poultry who seemed to go anywhere but the floor. The mixed mob in the yard outside consisted of fowls, ducks, guinea fowl , turkeys, geese and even an odd crow that had been known to have taken the wrong turning and entered the house with devastating effect on the room decorations and upper shelves. In all cases they took the opportunity of using the house as a lavatory. During our early years Mother had Judy, our spaniel dog, and several farm cats who used to keep the house generally clear of the rats, voles and mice that inhabited the buildings.. At least they did not nest in the rooms. Bats, long eared, pipistrelle and some larger horseshoe - lived in the roofs and caused considerable commotion if they entered the windows at dusk. My early memory was that any damage was caused by flailing arms, brushes and flapping clothing at their expulsion. It was some years before I realised what delightful creatures bats were, and I would peep outside the window to hear their squeaky voices in the gathering dusk..

Mother's early guests were her former colleagues from across the world.. We twins had to call all of them 'aunt'. Our favourite was Aunt Mac. A portly lady with ginger hair done up in a bun who Mother said came from Scotland and spent most of her life in the Far East.. She insisted that the hayrick visible from the living room window was Mount Fuji, that the cows were water buffaloes, the grain crops were rice and all high plants were bamboo. I found this very confusing but we got to love her rare visits as she always took us firstly to all the places normally forbidden by Mother.. She also allowed us to pee almost at any time ... as she explained that some native children in her part of the World never wore clothing below the waist to facilitate the act. A privilege not granted to grownups! although I remember inviting her to try.

Thus in fine weather we explored the forbidden Tip Hill, the wetter parts of Pool Copse, parts of the neighbouring Water Mead and even went one day to have a word with the Gypsies (Diddycoys) on the local Down. She would often ask Mother where we were not expected to go ... just to get new ideas. Perhaps her love of living dangerously led to her downfall for after many years Mother said she had disappeared.

A complete contrast was our godmother, Aunty Jo. A gaunt eagle eyed woman with a black shawl. We regarded her visits with trepidation. She was the wife of the London solicitor who had arranged for us all to come to the Farm. She could best be described as a holy terror. Luckily her husband, of similar ilk, had never stayed at the Farm because - so Aunt Jo pronounced - Mother would not keep a hot water bottle permanently on the lavatory seat, as he was said to suffer from a cold bottom. This seemed very strange to me at the time. but I hoped it was not catching.

As it affected me, Aunt Jo's crime was that she was a strong disciplinarian as far as small children were concerned . They should 'be seen and not heard', should never speak unless spoken to, they should at all times defer to the superiority of grownups. They should never get dirty, should speak clearly in proper English and should conform to very strict rules at table. They should go to bed early and never wake until required the next morning. They should only play silently .. if at all. Needless to say nature and 'modesty' had seen to it that she had no children of her own. We thought that her views were only exceeded by her husband who was said to think that the human species should not have an immature stage. Only the toilet seat had saved us! But Aunt Jo had money and a missionary zeal to go with it. At an early stage she must have considered that Dad was beyond redemption, after he fully displayed his table manner techniques on a large plate of soup while sitting right beside her. 'Lan' was expected after that to be eternally busy out on the farm and Mother saw to it that he never had occasion to explain to her his pig castration techniques in detail. Besides paying very well she arranged her own transport and brought a 'maid' to see to her every need. To ensure their due compliance the maid was a young girl who had recently had an illegitimate baby and , in accordance with the then custom, had to be banished to 'outer space' ... but our farm would do. The maid when not cleaning or helping Aunt Jo was expected to disappear. This usually meant trekking the remotest lanes and fields with her battered pram ... alone. We were certainly not allowed to in any way befriend the poor girl or share our secret places with her. Aunty Mac was never invited at the same time.... Alas!

At each of Aunt Jo's visits we would be given a revision course of table manners. In her long repertoire were :- never to sit down before she was seated, the fork was always to be used with prongs downwards, the eating irons to be rested between mouthfuls, the points never to be held above the handles, the soup bowl to be tilted away. The napkin drill was equally exacting. If she thought we had become slack she would glower at Mother and say "Beatrice !" in a penetrating voice. It was no good our crying, for immediate expulsion to the bedroom without food was the remedy. As a great 'treat' to us all. Aunt Jo would arrange her special brand of picnic. Her maid and Mother would prepare a somewhat elaborate cold 'dinner' in baskets. A taxi or hire car would pick us up at the farm and she, Mother and us twins would be transported to a beauty spot, often in the Quantock hills beyond Taunton. With her extremely limited respect for the powers of Nature she would pick some spot, usually under a great oak tree, and on a large white tablecloth lay out the banquet in detail with cushions suitably placed. During most expeditions we never actually got to the eating part at this stage for Mother Nature seems to have her own ways with the uninitiated . There was the dramatic occasion when a whole wasp colony objected to such activity near their nest. On another occasion thousands of long caterpillars suddenly descended on long silken threads upon everything like assailing shock troops. Nature also has its gas attacks; a growth of Stinkhorn toadstools decided to attract flies with their putrid smell and a large fox had decided to 'sleep the eternal sleep' just over the bank. This says nothing about the thunderstorms, high winds and sundry prowling dogs and feral animals that inhabited these hills. Aunt Jo's fury at being frustrated was good to see. Mother secretly admitted to us afterwards that the food tasted better as a real picnic in or around the waiting car where strict rules of etiquette did not apply. And it could mean another outing later! She had also been known to have left money for an outing AFTER she had gone.

Amongst our regular guests was a rather strange elderly man, he usually dressed in black, sometimes spoke to Mother in French, and seemed not to see infants. This was a relief for we did our best to ignore him. He ate his meals at odd times and would read most of the time sitting in the garden. or later listening to the wireless. Mother told us he was a retired parson who lived in France and used our farm as some form of a retreat. He seemed to regard Dad as just another part of the farming and as a weather forecaster. Dad was very good at forecasting the weather, just for the farm, of course. By the time we were about four years old - despite identical treatment - but not being identical twins -we already had begun to show differences in character . I preferred the life of the more feral child while Peter would prefer to stay around Mother and looking at books or playing games with cards or on boards. The divergence gradually came when we started building 'tree houses'. In fact he left most of the work for me These were usually simple structures probably in the manner of those made by the great apes, or perhaps baby apes. We used branches for the base, walls and roof , Long grass, straw, or bits of old linoleum made a slightly waterproof roof. We had usually four sites around the farm, Our very special one was in the 'Owl' tree in the corner of hind house. A natural platform ten feet up in an ancient polarded ash. Here we could see the house, the public road and the entrance track as well as distant woodland and even a neighbouring farm. Our second house was really a structure on top of the rim of an old chalk pit in 'home field'. We tried to roughly thatch this with old straw or grasses. Here we could see right across the valley and look down upon the whole homestead at the bottom of the hill,

Our third house was in the hedge along Tip Hill towards Fairyland. Here we could keep track of farming operations. This was low down in a knarled and ancient ash. Cattle could reach up to chew the thatch at times.
Our fourth house was on the border of Fairyland. A useful place to rest on journeys to our fields beyond.

In our very early days at the Farm there were no other young children near enough for us to mix with. By the time we were six in 1926 a house and petrol station had been built on the far side of the main road that bordered the entire top end of the farm. There were several children but Mother bid us not to play with them as we may cross the road . We did occasionally meet them while out blackberrying or picking mushrooms or nuts. But as Dad had yelled at them several times to keep out of our fields they never entered our most 'sacred' tree houses.

As we became older there was one regular visitor from Mother's governess days. She was apparently a widow, about Mother's age , and for some obtuse reason we were never expected to call her Aunty. She was just Annie and she had a daughter at just our age. We thought that she helped Mother in the house, and was a professional cook and housekeeper, she paid Mother very little. She had just one daughter Susie. She had been brought up without brothers and we had no sisters and being a friendly child she sort of joined us as equals ... we just wore different clothes. It was the tree houses that brought about the first problems.

We boys had long since given up returning to the indoor lavatory when nature called. So we found a tree house made an excellent urinating point provided no one was underneath and the wind direction was right. Susie used to watch us performing. But demanded to get down the tree and we were not to look as she hid behind some shrubbery.

This seemed to us both as a little unfair so she agreed to discuss her problem.
"You haven't got a ... a Willie. Have you?" I said
" I know I haven't " replied Susie in some irritation "Girls just haven't."
Being of an inventive mind Peter sought some way to seek a cure. So we agreed to watch her at close quarters as she tried to urinate from the tree top.

"Splashes all over the place like that .... can't you do something about it. " I observed.

" Like a storm of rain down below," was Peter's remark

It was the next day when we had managed to get Susie up the Owl tree that she found a solution. We had to do something for in this tree we had our main stock of lemonade and a tin of Andrew's Little Liver Salts and a bucket of water. On a warm day we drank a lot. There was also a hollow centre to the tree and we had managed to open a small shaft for Susie to use. So after carefully examining Susie's plumbing system we showed her a way of exposing her 'inner needs' by using both hands. We were now all equal .... almost. We found that if she allowed us to view her efforts first we were better able to direct our own jets clear of the tree. She told us her Mother used to examine. wash and powder her plumbing system daily before bed. After many years she and her Mother moved far away and tree houses were never quite the same afterwards.

During these formative years Mother had paid visits from both her relations and also Dad's. We twins did not like this as we were paraded, and the only family twins. We felt a bit like prize pigs or sheep. We had to wear 'sailor suits' or identical stiff shirt and short trousers. Mother would tell us not to play as we would get dirty and our hands would be inspected. It seemed that we had very few relations of our own age but a few young adults were allowed to take us for walks .... we on our 'fairy' cycles.

Over the years there was one very large London family whose younger members used to stay with us in pairs. One very young of about three or four and an older one of about ten to do the minding.

Like all Mother's young guests their parents would bring them down and collect them during school holidays. The parents staying the night on each occasion. This was the occasion of the great evening meal. Like most of the guests they relished the isolation. the home grown food and the primitive conditions. Dad had settled into the picture and they found his habits most engaging. Mother had long since given up worrying about his more explicit accounts of the basic sexual activities of farm livestock. Of course the end results ... the farm babies - both furred or feathered - were always popular.

It was indeed Mother's paying guest business that saw the farm through the deep depression of the early thirties. One result was however that in 1934 at the age of fourteen I had to leave school to replace Tom the last paid farm worker.

Now some of the quests were anxious to try out some of the more mundane work that I did such as muck 'spurring' when they realised that their own techniques often put the muck upon themselves rather than over the field.

I was to get no pay but I knew enough about the countryside by then to use any spare time I was granted to catch moles, stoats and weasels for their skins and have a few chickens to rear for egg production. After a few years I would be privileged to be the farm rabbit catcher - on a 50/50 basis ... in my spare time.

Chapter 12
OUR TRANSPORT

Certainly the Trap with Whinnie in the shafts was our earliest family transport. It was four miles to Chard and Crewkerne but as Mother was not confident with horses we were limited to when Dad was not busy. Which occasions were very rare. However the milk - at first one churn - had to be taken each day the three miles down the often steep lanes to the Milk Factory at Chaffcombe Dad would go on the extra mile to a shop, Abbots Store, near the Chard passenger Railway Station (GWR). This was once a week ... only if it was fine weather, of course. It was the one day a week that we had a newspaper .... Dad favoured the Daily Mail. This was subsequently used in the lavatory. As far as I recall other paper ... wrapping. paper bags or sundry used stationery was used unless Mother had guests

There were occasions when Dad used the wagon - with Smart in the shafts -to get cattle food or the odd building materials. Then it was the Chard Trading Company at he bottom of East street. I believe - except on hills - Whinnie , with her trace harness jangling - would have been tied at the back of the wagon by her halter. She was hooked in front for the hills.

Dad also had another perhaps more respectable sort of cart - called a governess cart - which was rarely used as it was lighter, with springs but was subject to structural failing ... due to the 'shackling' it received with Whinnie in top gear.
In any case Dad never wasted a journey on just simple pleasure.

Mother in our early days had our pram. With the rough tracks and the fact that mother did few journeys 'just for pleasure' but ensured that the pram was doubled up as a carrier of sticks, sundry wild fruits' and foodstuffs from Chillington Shop - the pram had a distinctly battered look. I am sure that there must have been one loose section secured by Dad with a piece of wire. I never knew what ultimately happened to it although in all probability it was parked under roosting poultry and was taken out with a load of sundry muck to 'fill a hole' somewhere around the farm.

Mother also had her 'one speed' bicycle with a front wicker basket and a carrier on the back. She would take us boys - one at a time- to Chard, perhaps to see the Doctor, or just to keep us apart in the interests of our 'minder' left at home. Mother must have been strong at the time for her front basket was always piled high. I shall always recall the straining noises given out by the old bike as she pedalled up most of the hills. I was soon realising that my parents never shared interests or abilities, so Dad could not ride a bicycle. When we were about five years old Mother got us tiny 'Fairy Cycles' and then we could ride on the flatter roads with her. The traffic was so infrequent, slow and noisy that even Mother considered it safe. And , of course, like ships at sea cars would often greet one another while passing ... but not lorries for they were driven by 'working class ' men for certain Then one day Dad purchased a very early Ford lorry. It arrived in the yard and a man was seen to get a lift in a car back on the road. It had solid rubber tyres on wooden wheels. It boasted a cab with a low glass partly-open front windscreen. It had only two pedals ... a clutch and foot brake. Two forward gears and reverse worked on a lever but the third top gear was on the hand brake'
It was, by construction, slow .... and Dad drove it even slower. Down steep hills it was probably unstoppable ... short of leaving the road or just rubbing the bank

Mother was considered the poor member of her family and although she had been ostracised by some members her sister suddenly decided, having heard of her plight in the backwoods, to GIVE her a brand new Austin Seven car. Not only that but an instructor would deliver the car and stay four days for Mother to learn to drive.
A few days later the car negotiated our track to the house and parked in the yard amid sundry hens , ducks and guinea fowl who promptly christened it during a close inspection. The instructor seemed a little out of place in these surroundings and bravely initiated Mother on some of the nearer flat fields before venturing out onto the roads. Apparently he never got round to teaching Mother to reverse so for a long time subsequently she relied on someone pushing her backwards when the occasion should arise.
 Being a man of some sagacity he never even attempted to instruct Dad. Mother told us later that the first time Dad took the car out he never managed to find second or third gear and returned from Chard over an hour later. Shouting 'WOA! when he fumbled for the brake in the yard and declaring that it was faster to take Whinnie and the trap next time.

Despite her reversing difficulties this car opened up Mother's world again. With a top speed of 35mph, down hill and often about half that around Windwhistle's steeper slopes. A month later she arranged a complete family outing to the sea.

She managed to persuade Dad to probably postpone a session of 'fencing ' or castrating pigs, put on his better clothes and leggings above his boots and make for Seaton, fifteen miles distant. We arrived at the promenade and parked the car, We all got out and Mother put an old blanket and a few cushions on the pebbly beach. Dad got out and went fast asleep as he had no interest in paddling with the rest of us. It was our first visit to the seaside and promptly at 1 o'clock dinner time we returned to wake Dad up and set out, on a tablecloth, a cold meal. Afterwards Dad immediately resumed his slumbers and Mother afterwards took us along the promenade to the junction of the River Axe and to see a few small boats.
All we had to do then was to return to the car, Mother christened her Jane, woke Dad up and returned to the farm. Dad never went to the sea again for as he said if there was no work to do he might just as well sleep at home.

Although Mother could drive Jane she never ventured under the bonnet and , although the car was extraordinarily reliable ... once it got started ... there was always a doubt that we would reach our destination.
.So Dad obtained a pair of Homing Pigeons, blue checkers they were called, and every longer journey we made meant that we carried a basket with a pigeon. On arrival the bird would be released and, at about one hour for twenty five miles, Dad would know that at least we arrived at our destination. The pigeons were normally reliable but we did occasionally get other peoples pigeons 'lodging' with ours for up to a week.

There was a garage in Chard - set up by a Mr Warren, who claimed that he was trained at Rolls Royce. Thankfully he did not despise such humble cars as Mother's 'Baby Austin'.

From the earliest days of the Baby Austin we did have one family outing each year. Dad felt the urge to return to his old house in the village of Bawdrip a few miles past Bridgwater at the beginning of the Polden Hills ... overlooking the Somerset Levels. Here his mother, a widow, was nearly bedridden and he would meet his two sisters. Both of whom were school teachers. It was around this house, spread in small parcels across the moors, that the fields that made up the original family small holding were made up.

Dad spent a whole year looking forward to this one day - between milkings , of course - of going back to his roots. The car was packed to the gunnels for, apart from the four of us, he carried some potatoes, a chicken, and a pack of our holly ... often tied outside the bonnet and due to the shaking on the roads minus most of its berries on arrival. We also had to carry the pigeon in its basket
Of course Mother came from a big house a mile away and although she could enter the house to see her brother and any other family member we had to stay outside

in the car. We never did understand what exactly was wrong with us. Just occasionally younger members of Mothers lot used to come and peer in the car, as if we were a couple of dogs. 'Oh! these are the twins", they would say to Mother with surprise ... as if they expected us to have long furry tails and whiskers.
Afterwards Mother would drive us back to Dad's lot. The tiny house was dark but cosy. Dad's spinster teacher sisters would have books for us to read .. or try to read. The garden was minute and we stayed indoors except on arrival when the ceremonial releasing of the pigeon took place.

The car was also in fairly regular use to drive us to the Private Chard School and collect us in the afternoon. That was unless Dad got us there , invariably late, in his very noisy and ancient lorry.

After three years the old Baby Austin was showing the effect of overloading on our farm track or poor roads. After a visit to Aunt Lilla and Uncle Henry they noticed the state of the car and promptly gave Mother a new hard top Baby Austin in brilliant kingfisher blue. Mother was never good at starting it with the handle so when out she often left it on a slope. Dad never liked driving a car as he apparently never completely got over the notion that it was not a horse.

When I was about twelve we bought an old Wolsey and then a Morris. A young farmers Club was formed in Chard and Mother occasionally, if it was too wet to cycle, used to drive me to the meetings. The Young Farmers Clubs were mainly educational and we had an Auctioneer as a Club Leader and a retired Butcher as a President. We had Club calves which we reared for a year and lectures and demonstrations. This was the time of the farming depression and there came a time when Mother was making more profit out of paying guests than was the farm. Although Peter kept on at School I had to leave, replace the last man and start life as a full-time unpaid farm labourer.

My saving was that I was allowed to make money off the land outside working hours. Mother used to drive me to more distant Young Farmer activities. The Young farmers were very active at that time and as chairman I had a lot to do .

But in the fullness of time the old lorry just ceased to function and milk was fetched by lorry. I first of all drove with Mother and then at 17 passed my test for my own licence. But I could not afford to have holidays and the only luxury was to attend the Art Class at Yeovil School of Art for five hours on Wednesdays in term time.

Although we used to sometimes hire a farm tractor and driver Dad did not approve of having a tractor of our own. Then the clouds of another War darkened the horizon.

I joined the TA Artillery at Taunton. They provided holidays, annual camp on Dartmoor, and petrol to attend drills weekly at Taunton.

Chapter 13
MILKING

Outside our back door was a concrete pad. Dad had rigged a simple galvanised sheeting roof over most of this. In cool weather this was the home of a couple of milk churns. A muslin (cheese cloth) strainer was secured for milk from a bucket to be poured in.. During warm weather a frame was used to hold a 'D' tank and a corrigated water milk cooler over the churn. As the tank was just a couple of inches higher than the roof it had to be open to the sky so that - in stormy weather - the milk was slightly diluted.

Dad only had about six cows in milk at first. The cowstall was opposite and here Dad -or a worker - would hand milk the cows ... sitting on a three-legged stool, The milker sat with his cloth cat turned backwards and a sack tied around his waist to act as an apron to catch the splashes as the milk went in the bucket -held between the knees.

Afterwards Dad required one complete kettle of boiling water each morning to clean the cooler and buckets. Owing to the somewhat primitive nature of hygiene at the time, the milk had to be taken to the Milk Factory at Chaffcombe as soon as Dad had had breakfast. Most farm workers could milk and one other person would help. We dipped out a large jug full ourselves and a small jug of cream would be taken from the settled nights milk before the churns were labelled.

Dad would always take the milk himself. Probably as Whinnie was a little excitable and he enjoyed the three mile journey with the trap. If Whinnie was in a field she had to be caught and this was a little tricky and could be a little time consuming.

For me fine weather meant accompanying Dad for the journey. Peter could also come but would usually prefer to stay with Mother around the house. The journey started off up Dad's stony track to meet the public road at Lodge Copse. Under the

Landing Gently by Paul Smith

Mountains of Muck.

'Spurring'

Where there's livestock there's always deadstock.

First catch your horse

Feeding pigs is not easy

A bit on the side

THE MOWING.

HAYMAKING PICNIC.

LOADING.

HIGH SPEED TEDDING.

RAKING UP.

HAY MEADOW.

LANDING GENTLY   *by* PAUL SMITH

HIGH SPEED TRAPPER

THE FIRST ROUND

NEARLY DOWN TO THE 'COMFORTS'

DON'T FORGET US!

THE ONE THAT GOT AWAY

LANDING GENTLY *by* PAUL SMITH

trees here we would sometimes see a red squirrel before joining the main road at the old gateway .. now removed .. after Dad had put up a barbed wire fence along the field.

It was always breezier along the main road for the mile under the great avenue of beech trees. Usually here we could see down over a great part of Somerset to the Bristol Channel ... a study of fading greens, blues and purples. Shortly after the Windwhistle Inn - with perhaps a wave from landlord Willy or Mrs Edwards we would enter a long winding lane with high banked hedges.

Dad seemed happier in lanes and would gently hum to himself .. no particular tune .. just a sort of noise that spelled contentment ... like a cat purring. We would very rarely meet other traffic and Whinnie would just keep up a steady trot on level ground. Without brakes she had later to sit in her harness down the steeper hills

At a staggered cross-roads we would then enter a very steep lane into Chaffcombe Woods with often banks of bluebells, foxgloves and a wealth of bright flowers under an oak canopy. Dad would often stop under a large cherry tree that spanned the road so we could better see a squirrel. or even the sight of the occasional deer or a fox.
There was a lone cottage in the woods that Dad called Stud.

Soon we entered the village and on past the War Memorial and the stone cottages each with a patch of vegetable garden and a few roses around the door. There was a gateway to the Big House .and the winding road out towards a glimpse of water through a hedge. This was the Chard Reservoir originally built to water the Chard - Taunton ship canal but it later was killed by the Railway in the last century. (1868)

Across the shaded road was the entrance to a redbrick building and the sound of clanking metal and the noise of other carts and wagons moving around. This was the Milk Factory. We waited our turn to reach the unloading platform .
 Here our churns were checked and emptied into a large vat. We moved further on to pick up two fresh cleaned churns. What intrigued me at the factory was how one man could manage to roll two churns, on their rims, at once across the platform . In the local farming society Dad was a still 'voreigner' on first moving in ...forty miles was a long distance in those far off days.

On one day a week, after the milk churns, we journeyed on the extra two miles to the outskirts of Chard to a small shop near the Great Western Passenger Railway Station.. Here Dad would pick up two copies of the Daily Mail and, perhaps, a small item of groceries - usually a packet of tea, sugar or custard powder - before retracing our journey. Whinnie was often sweating profusely but she was not allowed to drink in the Fore Street horse trough.

The journey was in parts steep and Dad stopped at Stud to hand a small grey-haired thin man his Daily Mail and remark on the weather in the best of local accents. The round journey would take some two hours and by the time we returned I had lost interest and would sit in the back of the trap on a couple of sacks and a small cushion.

After a couple of years the factory at Chaffcome closed down and the Milk Factory was transferred to Salter & Stokes at Chard Junction.. This was over five miles and Dad soon got his ancient Ford lorry.. The lorry was used at times to take us to school, usually ten minutes late, so this vehicle was held in some awe by Peter and me and, detested by the School as letting the tone badly slip.

But the great joy was that the Factory was just the other side of the railway level crossing at Chard Junction Station . This was the main London to Exeter line and when the up train stopped in the station the engine would stop on the crossing.

Great Engines like the PENDRAGON would belch out steam and apparently fume with anger at having to wait. The loud squeaking hoot and the grunt on restarting would invariably be accompanied by a screech of slipping wheels. The lorry seemed tiny beside such a brute. It was very exciting..

Sometimes goods trains would use the same line and would not stop. The engines used seemed long and straggly to me . This was the Southern Railway and the terminus for the Great Western trains from Taunton.. On this trip Dad would come back through Chard to the Somerset Trading Company and Bradfords to collect great slabs of cotton cake for the cows, milled grains for the pigs and the odd building supplies.

The journey back up into the hills was steep and the old lorry would sound as if it were slowly grinding to death up the steeper bits. I could not hear if Dad was humming to himself in the lorry but I expect he occasionally shouted commands as if it were a horse. In a few years the milk churns were collected on our stand alongside the main road at Windwhistle.

This was a time when we first saw buses, delivery vans and larger lorries. Even the postman had a bicycle. Dad's lorry was left in the rick yard where it slowly disappeared beneath nettles, brambles and a couple of elder bushes. The cab had been used for conversion into one of Dad's hen houses. The milk was taken out in the putt.

Chapter 14
LOCAL WOODLAND

Mother jokingly told us from an early age that our distant ancestors lived in the trees. This was despite the story of Adam and Eve and the fact that the human race, according to the Bible, had no ancestors ... they were created 'In Situ'. I now believe she developed this story to shame me into remaining at ground level.

From an early age I liked trees ... I regarded them as friends ... and felt an impelling desire to climb them. To be up in their branches was to enter another world ... truly to join them and to revel in the secrecy and safety that they provided. . At first Mother would restrain me - purely as it was dangerous. Peter never had this urge and Mother had finally resigned to the fact we were 'different'. She revealed that Dad had climbed trees as a boy. and that I had probably inherited his 'problem'.

I started with small elder, a few low cut slanting ashes and one-time coppiced nut bushes. Each season I managed to get higher and explore more of this wonderful world above ground. The area was peculiar to birds, squirrels and - I discovered- to a few other small boys. One boy at school had the same disease but had been in hospital three times with broken bones. His dad was not probably so sure footed.

There were trees surrounding the farmstead and in every hedge on the farm. There were the wonders of Fairyland and the Weir. But my great joy was to go beyond the farm boundary in the other woodland that existed down the valley. Chief amongst these were Eighteen Acres Copse. at the head of the valley and Pool Copse below the Farmhouse in the valley bottom . Here was my great escape ... from the age of about ten.

Pool Copse had a stream running through its centre with a pond - behind a retaining wall - at its lower end. This pond was the abode of thousand of frogs, toads and newts in the breeding season. Eels, some large and silvery, inhabited its soft mud floor. Wild duck - mallard and teal - regularly used its waters and moorhens mingled with the occasional kingfisher. There were great patches of bright yellow marsh marigolds and areas of golden saxifrage along the stream. An area of bull rushes and yellow flags had colonised one bank.

The floor was criss-crossed with paths of badgers from several nearby setts and roe deer and foxes often laid up in the denser undergrowth. I explored it with endless delight. .

But It was the great trees that were the true glory of Pool Copse. They rose up to a dense canopy almost a hundred feet above the floor. and a few had been blown against their neighbours so that their roots tipped out of the soil forming nesting sites for small birds such as wrens. There was a rookery in the high spruce and larch trees adjoined our old orchard.

The trees must first have been planted way back in the seventeenth century for some were very old. There were a few great ash trees, a huge towering sweet chestnut, two horse chestnuts on one border, two black poplars and a few oaks. But the great glory were the spruce, the feathery larch and the Scots pine.

There were two places where I could climb up into the canopy. A leaning larch enabled me to clamber up half way up a spruce where there were branches to enable me to reach the canopy and go from tree to tree. This route had its drawbacks. First the spruce trees oozed turpentine from their upper branches and the rooks claimed this spot for their rookery and in the nesting season - from February until June - noisily drew attention to my presence and often did not hesitate to deposit their anger upon my bared head. Due to the turpentine I became sticky and my hands would take up the green algae so that. Mother did nor approve.

Another route was at the far end and this time up a leaning spruce into poplar, then some larch for about a hundred yards to where I could descend an oak ... with difficulty. There were several places where I could sit quietly and look out over fields, the road , a neighbouring farm and views of bits of the old orchard and the weir. It was very important to get down to the ground again in good daylight.

On one occasion I had climbed an oak tree overlooking a badger's sett. At dusk I saw the shadowy movements of the badgers as they went for a drink before making off into the depths of the wood. It then became quite dark as I had to descend. I nearly got to the ground when my trousers hooked in a branch and in trying to release myself I slipped and landed up part naked on the soft earth of the sett. As I lay there a large badger came right up to me. I managed to clamber up when the moon came out to retrieve my torn trousers. Mother was, needless to say, not pleased.

The pond had a sluice with piping under the road for the stream to be continued through a hundred yards to piping which took it half a mile to the pressure tank of the great hydraulic ram that supplied Cricket St Thomas House and Home F arm. This hundred yards of stream was in woodland. For some reason the trees here were mostly pine and quite unclimbable. From our earliest time at the farm mother would walk with us along the path beside the stream. We would both have little boats - made of wood and highly painted and which would go at slow walking speed.

As the main paths through these copses were walked by the Estate game keeper most days Mother insisted that we keep to the path so as not to disturb the pheasant population .

The other main woodland of my childhood was Eighteen Acres Copse (it was in fact only barely fourteen acres). This was a totally dry wood mostly on chalk. It was an old shooting place with three rides and a surrounding passage way. Here the trees had mostly been cut down in the Great War. The exception was a area of old oak trees in the centre which contained a large rookery.

About an acre in its heart was also coppiced hazel, and a hurdle maker would work here in the warmer months. We only saw him at a distance and Mother told me never to enter the wood if I heard him chopping. I thought he must have arrived on foot. His completed hurdles were laid in a neat pile awaiting collection by horse and cart from the next farm. There was his thick flat base board with the holes for the vertical shafts and a pile of hook shavings.

The contrast of this wood was greatest in the plants. There were swards of primroses along the rides and the bluebells before the shade became intense scented the air and supplied playgrounds for badgers. There were other unusual flowers, unnamed by me at that time - such as twayblade, herb paris, butterfly orchids, enchanter's nightshade, bugle, wild garlic and many others . There were blackberries and nuts to pick up.
In parts the vegetation was so dense that I would lose my way...and only gain my bearings by going downhill as the whole wood was sloping towards the west. Only I, in our family ever entered this wood and I was always nervous that I would meet strangers, dogs on the hunt or, worst of all, a pack of fox hounds in full cry.

There were only a few low bushes that I could possibly climb to safety. Some of these were coppiced hornbeam. I could usually detect if the hounds were about because the badger holes would have been roughly stopped with bits of wood and a few spits of earth.

So I would usually use Eighteen Acres Copse as a short cut to get to Fairyland, the next wood down the valley. There was a corner of our neighbours field that was a glorious spot for yellow trefoils and clovers. Here I would be fascinated by the numbers of blue butterflies. I could enter our own land in several places if a saw anyone or heard noises.

There was other woodland in the locality but as a small boy I had learnt from the wild creatures to always be wary of strange people or even domestic animals .... they were all much bigger than I was! I had learnt to walk stealthily under trees carefully avoiding dry twigs that would crack under my weight and startle wild creatures. I had to avoid walking on the skyline and learnt to stand in front of trees so as not to reveal my outline so easily.

Dad had shown me how to make bows our of ash suckers with arrows made of dry nettles topped with elder. I also made whistles out of horse chestnut twigs as the sap would rise in the spring. I could use tough string as a sling for a hand thrown Whitby spear. As I grew older I was able to adapt many of these ideas to become a real hunter for bringing back our dinner. .

Chapter 15
'ITEM'

Mother had always kept us twins on a fairly 'tight rein'. But as we gradually developed in different directions she tended to look after the one that followed her most closely and - like the twin lamb that gets eaten by the fox - I often strayed to the wilds of the farm.: scientists would have said I had a tendency to go feral. Dad was far too busy to watch me I often returned from my adventures with bruises and sprains and had to endure the stinging from a liberal application of 'Horse Embrocation, it would remind me to be more careful in the future. I had also learnt to judiciously apply a cobweb to a bleeding scratch.

Apart from the usual boyhood sins like the dislike of washing behind the ears and wearing out clothes and footgear , I cannot recall that we twins were ever the cause of real mayhem. But in the fulness of time Mother was to import a paying guest whose behaviour would rival the exploits of William in Richmal Crompton,s 'Just William' books.

For some reason that escaped me at the time all Mother's Paying Guests seemed to love the primitive wild isolation of the farm. They also delighted in the unorthodox agricultural exploits of Dad. So after a time Mother had a waiting list of guests .. especially for the school holiday period.

There was one large town based family whose mother would produce a new baby with great regularity and for the fortnight of her confinement she would park an assortment of her previous brood on our farm. Thus we would regularly have a young lad we called 'Item' (I cannot recall why!) and his two elder sisters to 'mind' him. Sometimes they would have a younger toddler as well.

Item was a very intelligent, active and inventive boy , who Mother inadvisedly trusted, and who actually devoted most of his abilities in tormenting his older sisters - and in so doing causing alarm and mayhem,

One particular acquired skill of his was the catching of frogs and toads and the occasional newt. His sisters were particularily sensitive, well mannered quiet girls and with splendid regularity at least one would be heard to rush screaming, having suddenly found a local amphibian or a large furry spider amongst her belongings. Or even her wash basin or the bath. Item was never caught in the act .
Item would supposedly make amends by presenting them with a basket of mushrooms, nuts or even blackberries. I would await the scream as they would rush to Mother , myself or on rare occasions Dad - to rescue them from certain death. For amongst the goodies lurked a 'baddie'.

Of course some of these beasts would be seen to escape in the house and the search party would rummage amonst wardrobes, chest of drawers and cupboards with Item supposedly 'helping'.

More than once when his sisters had put him in charge of the Toddler -an infant of some two years old- he would arrange for this also to escape, usually into a dangerous situation amongst cattle, pigs, poultry or even horses. I am armazed - even to this day - how livestock usually do not harm a 'man-child'. But just give the grown-ups appoplexy. He once allowed the toddler to sink up to its waist in soft manure on a tour of the cattle yard. .

It was always very difficult to pin blame on Item but to us it was very noticeable that in his absence wild creatures kept their distance .and farm stock - more usually- kept their places.

On one occasion when Mother had prepared a great evening feast of jugged hare and the table was full. Item had been given an early dinner with us and his sisters. Dad had gone out to put poultry, ducks , ferrets and calves to bed . Suddenly a commotion in the garden revealed a sow and piglets playing in the cabbage patch. Item was first to help to try to get them out of the garden.

Of course his sisters were old enough to deal with him ... that is what they were there for ... but they lacked a vicious streak and had to endure the pranks. Their hats, left on a tree trunk while they were playing were mysteriously nailed to the bark on their return. Their slippers often 'collected' small stones in the toes and they even suffered 'apple-pie' beds. Indeed one 'lost' slipper turned up twenty years later.

One of the most hurtful tricks he once played on the older girl. when she was about sixteen years old, and was well planned. They all had pocket money supplied by their parents. But on this occasion she ran low and Item 'gallantly' volunteered to lend her some of his. He required her to sign a note acknowledging the loan. The note however had a cutout, almost invisible section, for the signature. This enabled him to have a blank sheet of paper with her signature at the bottom.

He then wrote a passionate love letter, addressed to a neighbouring farmer's son, and putting in an unsealed envelope delivered it in person to the mother of the boy - who was already engaged to be married. This lady in due course stormed over to our farm to see Mother and to know what went on. Amid the ensuing puzzlement and tears Item had wisely persuaded me to go with him for a long country walk!.

Item was a regular guest at the farm over many years and never seemed to mend his ways. He was my first meeting with a lovable rogue.

Chapter 16
HARD LABOUR

I was not happy at Chard School. It was really a boarding school for pupils of more affluent parents whose fathers were in the Armed Forces , the diplomatic services or just living abroad. The school.activities took up all the time with games to finish the afternoons, Mother would not let us do games as we had five miles of hills to get up by bike at the end of lessons each day.

Farming was in a deep depression and the mortgage had not recently been paid.. Dad and Mother had to take some drastic action. The farm could no longer afford any paid labour and they might have to sell up and Dad get a farm job. I had been offered a job as a very junior office boy in the London solicitor's office of Aunt Jo's husband, Peter would also have to leave school for a job.

There was one solution. If I agreed to leave school and replace old Fred, our last labourer, and receive no wages Dad said Peter could remain at School for another year as he so hated farming. and we could remain in the Farm.
I would work every day and get no direct pay. But I would be able to devote any free time to 'living off the land' and making pocket money.

Mother had, up till now insisted that neither of us twins did regular 'farm work'.I suppose that where she had lived 'labouring' was for the 'natives' and she did not realise that I had inherited Dad's belief in 'hands on work'. He had been compelled to leave school at eleven to take over from his deceased father - my grandad.

So I really had no option as I would lodge in London with Aunt Jo in a house whose sole comfort was a warm lavatory seat. I had never worked in an office. By the age of fourteen I had grown to love the lonely wild conditions of our farm life.

So on my first day Dad roused me at 6 O'clock and in the November darkness we lit two candles and he used one match to light a piece of paper in some tinder - larch bark and dry sticks - in the kitchen stove. While I waited for the kettle to boil he quickly looked around outside to see that no disaster had struck in the hours of darkness.

Putting on our boots and old clothes we quickly had a cup of tea .almost in silence; and by the light of Dad's old hurricane lantern made our way out to the cow-stall. As we only had one light I just followed him around and awaited his instructions. The darkness was a bit frightening as I was never up at that early hour.
There was a concrete slab outside the back door and on this was placed the stand for the milk cooler -- the milk being poured in the 'D' tank above. But this morning the warm milk from the buckets would be poured directly into the muslin sieve at the neck of the churn. The frost would cool the milk sufficiently ....apparently.

Dad told me how to use a pick (fork) and put a small quantity of hay from the forebay into each cow.. More hay being fed to each cow as it came to be hand milked by Dad sitting on a little 3-legged milking stool. into an ordinary galvanised iron bucket.

He had two buckets - which he used alternately - and after each cow was milked I had to take the milk out to the churn outside the back door and pour it into the muslin strainer. As the buckets were open topped I was surprised at the debris that fell off the cow or the bits of straw that wafted into the churn . As both churn and cooler were open to the skies a thunder storm or heavy rain would certainly increase the volume of milk.

There was usually one churn for morning milk and another for the evening milk. It used to take Dad about ten minutes a cow and so, until I gradually became efficient we had a late breakfast.
My initiation into milking was painful, for cows not only have back feet to kick a small boy precariously balanced on a three legged stool with a bucket between his knees. The front foot can also reach back to kick up bedding in my face . After a cow has laid down it often lets its tail trail in the gutter so that this muck-laden wet tail can be used to give a small boy a sound whack around the ears. The art of milking also requires that the hand symmetrically squeezes from the top of the teat as well as pulling down.. At the end of a week my hands and wrists were aching and swollen. Dad's cure for such things was Horse Embrocation so my hands also suffered from dermatitis.

After a very quick breakfast the stalls had to be cleaned out with a curved dung fork. This was while Dad caught Whinnie and harnessed her ready to take the milk either to Chaffcombe or to the milk stand half a mile to the main road for collection by the churn lorry from Chard Junction milk Factory. Dad's old Ford lorry had to be abandoned in the yard and was now shielding a patch of nettles and thistles.

There were usually a few baby calves either to let loose with their mothers after they had been partially milked by hand. or fed out of a bucket by sucking two fingers of a hand placed under the milk.

There were a few older young cattle who had to be given hay in a rack. The two horses were sometimes in and had to be fed. Any animals kept in had to be mucked out and straw used to re-bed them .

The most important biological fact I learnt in those first few months was that more comes out of farm stock at one end than goes in at the other. So 'mucking out' was a daily chore. The muck was put into stacks or a central heap and each shed or byre seemed to me to have a mountain of muck beside it. Here it could be hand loaded into the putt and hauled to the fields where it was tipped into 'wheel-barrow' sized heaps in lines about eight paces apart across the fields...

From here it had to be 'spurred' (spread) with a long muckfork. There was skill in spurring and it was necessary to 'flick' the fork, holding the muck; at the right moment to get a spread that was 'zuent'(even) and, as so many beginners discover ; not to get any in the hair or down the neck of the operator. .

During the winter months getting the cattle food ready was a regular job. The mangel-worzels were stored at the end of the barn and they had to be put through the mangel slicer and put into a 'willy' basket to take to the stalls. They were fed by hand in the top and a large wheel turned by hand.... my hand - of course.
The cattle were fed hay from the ricks during the winter. This was cut into square yard slabs by a hay-knife in the hayrick. Dad would then carry these slabs ... about a foot thick, on his head to the stalls. I found this knife difficult and heavy work to handle for a time. It was possible to amputate one's toes in an uncautious moment.
Any minor scratches could have a cobweb applied to stop bleeding . This was one of Dad's miraculous cures. I was always a bit wary about complaining of muscle pain or sprains as horse embrocation would cause painful skin trouble.

The cows also had great slabs of 'cotton seed cake'.These had to be crushed before being fed, through a cake crusher ... again turned by hand.. The third barn instrument was the chaff cutter. The hay was sometimes fed to the horse at work and when this happened it had to be cut to about one inch lengths by this machine again turned by hand as the hay was fed in at the other end. ;

Of course there was the poultry to feed, unless chicken or ducklings were being reared in pens, grain was just chucked out on the corner of the rickyard. Dad used to like to feed the ducks just after dark .... he said that would ensure that they could be penned up away from the attentions of foxes. Then he had a pair of ferrets which he fed on bread and milk unless they were being worked when they had rabbit liver or guts. The larger dog ferret for some reason used to live in a rusty ancient 17 gallon milk churn partly filled with hay. ..

Each year a hedge would be cut and laid ... with some banking if time allowed. The only tools used were an axe and a bill hook. After the heavier wood was cut out for fencing bars or posts the brushwood was tied in faggots for use around buildings , under calf bedding or put in a stack in the rickyard. Beside this was a 'wood horse' a sort of chopping block. These faggots had to be cut up into one foot lengths with a bill hook for the kitchen range or the sitting room fire. There was now no coal . and fires burnt a lot of small wood.

In wet weather there was always the yard to scrape and the drainage ditch in the field to keep clear. Keeping the hedges cattle proof was a constant job. Dad's method was to cut thorns or brambles and picking them up with a long fork to stuff them in the hole. I used to get so many scratches that I just had to hope none would bleed too badly. Cobwebs were not readily available in the fields.

I always hated certain jobs with the pigs' All adult sows out at grass had to have six metal rings squeezed into the sensitive rim of the snout. The sow had a loop of a rope slipped into her mouth and the noose pulled tight, the end was tied to a bar or post and the sow would hang back squealing in apparent agony, ringing pliers were then used to insert the rings, one at a time. The rope was then released and the noose slackened.
Worse still was castrating (cutting) the boar pigs at three weeks old. In this operation I had to hold the piglet by its back legs against my knees while Dad cut the scrotum with his pocket knife and squeezed out the testicles with the piglet screaming 'blue murder'. The testicles were fed to the ferrets or the dogs. While this was going on the mother sow would be just outside the door waiting to tear me to pieces if I stepped outside.

Another tricky job I was given was to catch Whinnie to harness her to take up the milk. Whinnie was usually in the two steep fields below Tip Hill . I had to tempt her with a few knobs of cow cake in a bowl .. to offer a hand to Whinnie could mean getting bitten. for the unskilled. She would not allow me to carry a halter .. at least not showing. She would wait one side of a field and when I was within a few yards gallop to the opposite side. When I got half way to this hedge she would run back kicking her legs up in the air as she passed me.

After a time I discovered a method which, for some reason, seemed to outwit her. I would walk obliquely to her pretending to be looking for mushrooms. She would watch me until I got within striking distance when I would grab her forelock. As she tried to shake me off her momentum would enable me to get on her shoulders with my arms around her neck .. for some reason she would respond to a heel being kicked into her ribs as I steered her to a gate and a halter.

She did have one final trick. One hedge was badly overhanging and she would run under the overhanging branches hoping to drag me off. .... by lying quite flat and with my head well down I was just able to get under even if I got a bruised shoulder or bottom. Dad would not allow me to use Flower who never resisted being caught.

There was always more to do than I could possibly get through in any one day .This is not to mention stone picking, thistle cutting, or seasonal work at haymaking , harvesting or crop gathering.

I think Dad would have liked me to work a twelve hour day and he often criticised my work But things were going badly for farming and to greatly add to our troubles I had barely got into the routine when nineteen pigs - of all ages - were poisoned by some paint thrown over the main road hedge into Eleven Acres. They had to be buried and so I spent many days as a grave digger. Any cows that died had also to be buried and I spent much ' spare' time digging into the chalkier parts of the farm.

During my first year I was often very tired as the work was very hard and continuous over a seven day week. But I was finding some spells of spare time. I had learnt to catch moles, and a few stoats and weasels for their skins. I had gathered mushrooms, nuts , sloes and crab apples for sale and had bought a few hens of my own I had a calf from the newly formed Young Framers Club.

In the evenings after dark I would write a few articles and short stories for pleasure I also did some drawings of wild flowers.

At the age of fifteen I won a national essay competition run by the 'Young Farmer Magazine'. The then editor also agreed to print some of my cartoons. I was allowed to own my own dog and I started breeding ferrets. I had become a workaholic but I was carefree. While Mother encouraged me in everything I did , Dad had lost a great deal of his zest and seemed at times despondent that the farm was not fulfilling his dreams.

As the years went by farming began to regain some prosperity. Mother always encouraged me to look beyond the boundary fence. In early 1939 when I was nineteen I perhaps carried that advice too far. .

Chapter 17
HAYMAKING

As soon as the first signs of spring appeared all the cattle except the small calves went out to grass in galloping frenzied excitement. Dad's policy was simplicity itself. While the cattle - and horses - had the steeper, lower and nearer fields the other half... about sixty acres ... was let up for mowing. Mother duly warned us never to tread in the mowing grass ... at the peril of Dad's wrath. So we had to judiciously mingle with the livestock; to many the sight of small children and a small dog acted as a magnet.

While Dad saw to it that all the winter quarters were duly mucked out, Mother caused turmoil in the house with Spring Cleaning. She used to take a room at a time. All moveable items would be taken to the garden and parked on the grass. Carpets and matting would be draped on the washing line. The main floors were flagstones and these had to be washed. Then there were all the carpets and rugs to be beaten and Mother had a special antique wickerwork carpet beater. Often clouds of dust .like smoke from a forest fire wafted out over Tip Hill. Perhaps we should have been thankful that we had no near neighbours. On these momentous occasions small children were not welcome.. The rooms were done singly and as children we never knew which would be next, so there was an air of dreaded expectancy as Mother did a bit of what she called 'cogitating'.

These early months were spent on maintaining the fences and seeing the gates were shut. In times of slow growing the animals spent much of their spare time proving that the fences could be breached but Dad trained his dogs to chase away any faces that looked through the hedges. ... this was done by Dad with the special shout 'skiffurum' and blowing his whistle when the animal had again regained the centre of the field. In fact Mother soon christened dogs 'Skittybeasts'.(We use the word to this day!).

Many of the fields had, many years ago, been marled - as shown by the pits in

some fields they would have been a mixture of chalk and clay.
Now with no modern fertilisers nature provided its own nitrogen in the form of clovers. In poor parts of a field yellow rattle fed on the clovers . By the time the grass was bulky enough to cut, the hay fields were a mass of flowers, butterflies and bees. I used to lie on the edge of a field and looking just over the level of the grass see the insects rising and falling as if performing a dance. The scent of a newly cut field was sweet and highly aromatic.

Dad's old two-horse mower had been handed down to him by his father and tended to rattle in every joint . The two horses, Flower and Whinnie, walked either side of the centre pole attached to their collars. Not a good match of course for while restraining Whinnie he had, at the same time, to encourage Flower. The mower, with Whinnie on the outside, tended to come out of the grass. With these early reciprocating mowers a constant hazard was mouse nests. These would catch on the fingers and it would be necessary to back a foot out of the crop, lift the knife, put it out of gear and knock the nest off the finger with a stick Dad carried. Hoping all the while of course that Whinnie would stand still.

Every half hour the knife would need sharpening and so Dad had two knives which he held on the top bar of a gate with a long gate-vice and sharpened them with a flat 'whetstone'. When Dad was working in the far fields Mother, when we were quite small, set up a small camp area under a tree. She would then light a fire and make tea with a kettle or a cauldron held over the fire in the fork of a stick.

I don't recall her bringing saucers so Dad had to cool his tea by adding extra milk. There was one great problem for the horses were constantly twitching their heads and tails and stamping their feet with the flies. Some of these were the large horse-fly which had a vicious beak which it stuck into both humans and animals. When he had his tea he parked the horses nearby. Mother let us twins move into the smoke from the fire to keep the flies away, and which was allowed to slowly go out after the tea pot was filled.

When a field had been cut it was usually next day before Dad used the swath-turner. This was another ancient rattling machine with revolving disks like large metal holly leaves. With this the swaths were turned over to dry the other side. Whinnie usually did this job but if she was allowed to go too fast the grass would be sprayed in the air and the swath destroyed.

After two more days Dad used a one-horse hay rake which could put the hay in rows across the field. These rows were again raked into blobs of hay ... ready to be picked up into the wagon.

READY FOR A BUMPY RIDE.

Landing Gently — by Paul Smith

CONCEALED ABOUT THE PERSON!!

BEDTIME

"TROUBLE IS.....'E CAN'T READ"

MIDNIGHT HOUR EARTH STOPPING!

FIT FOR A KING!

LANDING GENTLY — by PAUL SMITH

Dad had two wagons, the second one had a bit of floor board missing and one back wheel had a rotten spoke .. It was used as a roost for fowls. The good wagon was a very solid affair. Heavy shafts with a back platform for the driver. The wooden sides had lips with metal grills. These Dad referred to as the 'comforts', for when unloading hay, straw or sheaves it was a great comfort to have a solid surface on which to stand for the finishing bout.
Dad insisted that all crops were brought back to the rickyard ... a relic of his days on the wetlands. The wagon was quite heavy to pull by itself and as one bit of the journey was uphill Dad used Whinnie as a trace horse. She wore special pulling harness with two chains attached to the front of the shafts. The problem here was Whinnie's enthusiasm and on corners or uneven ground the load was apt to either slip or even tip on its side ... which was, to say the least, very uncomfortable for Flower in the shafts.
Apart from this the whole load had to be shifted clear of the wagon and Whinnie, with her trace harness linked to a rope used to right the wagon.

The journey from the rick yard out to the hay field was a very bumpy ride but, while Dad sat on the shaft end holding the reins, the three men would sit on the comforts but we children, as a great treat, would be allowed to sit on the floor of the wagon for safety. When the wagon was empty the eight foot lades would be unhooked from the floor and would rest on the middle section of the comforts.

A long rope, known as the wagon line, was folded by a special way using both arms and this would lie on the floor. In addition were three long pitching forks and two short loading forks (for some reason forks were known as picks around Windwhistle area). Outside on a tough chain attached to the wagon body was the drag shoe. This slid under the back wheel and was used as a brake when going down hills. Dad had fixed a second brake shoe to the other side for steep hills.

There were no springs whatever on the wagon and apart from getting shaken up after the first unloading a large menagerie of creepy crawlies occupied the floor space. There were long legged harvesters. daddy long legs, the devil's coach horse beetle, various caterpillars, woodlice, smaller spiders, shield bugs and even some unidentified small denizens of a hay field. As the wagon continued its bumpy journey these creatures vainly tried to climb the trembling sides and any rising surface, such as a small human body, that presented itself.

Although I enjoyed the ride Peter usually decided to stay with Mother. As the weather would be sunny he would read books or sit on a blanket in the ease and security of the garden. I would join Dad's world - a mans world- out on the farm.

On arrival at the field the outfit would be directed into the first double row of hay cocks. The rope would be hung on the back of the wagon and the lades would be erected into position. Dad and one man would take the long pitching picks and pass the hay to the third member who would take it in and make the load in layers. With Whinnie in the trace harness and Flower in the shafts both animals would respond - most of the time -to Dad's commands of 'gee-up!' ,'wook-off!' (right) or 'come-to!' (left) or 'wooah' (stop) . I noticed Dad did has an alternative set of clicking noises if he was standing near the horses. It usually happened that Whinnie and Flower responded differently so that they had to be straightened out at intervals.

When the wagon was above the lades and the load maker appeared in danger of toppling off the top, Dad would call a halt and the eye on one end of the 'wagon line' secured in the hook under the front uphill comfort. The rest of the line would be thrown over and secured under the hook on the other side. With one man keeping tension on he end it would be pulled tight.
Again diagonally across the load to the back .... again in the hook and then over for the final hook. When all were again pulled tight the loose end would be passed to the ropes to give extra tightness.
When Dad was happy the picks would be stuck in the back and the tortuous journey back to the rick yard begin. When I was about seven years old I and one of the men would scramble up the rope to the top and he would hold on to me for the hair-raising journey .... at one point in Home Field the wagon would pass near a pollarded ash so that the branches scraped the top. Here we had to lay flat and tightly grip the rope.

Back at the rick yard the wagon was drawn alongside the rick - usually one per field - and the line removed. The ricks were made square on a bed of wood faggots. These acted as a damp course but also a home for rats, mice and other small beasties ... even grass snakes. The rick bottoms were a good place to exercise ferrets ... so I was to learn later.
As the loads were laboriously unloaded with a pitchfork, the second man took it in and passed it to the rickmaker .... who went round and round also keeping a slightly higher centre.
The last two loads would be used to form the roof. The rick was then raked to remove loose material and thatched ready for the winter.

There were rainy periods during haymaking and the hay had to be made into haycocks in the field to run most of the rain off. But my memory, as is common in old age, excludes the bad bits. For me haymaking was a time of perpetual sunshine and excitement. Mother's guests would also share in the fun.

At first only the muck from the livestock was used to fertilise the hay fields but gradually ground phosphate, phosphates and bagged nitrogen were applied more frequently as each year went by and I was the labourer who did much of this work.

Chapter 18
HARVESTING

Although much of the farm had been sown to corn before we arrived, Dad was a grassland farmer and my early memories of harvesting were very primitive. In fact for a few years he only grew a few acres of wheat, oats -or once rye- to feed the poultry, supply some bedding for winter stock and provide reed for thatching..

In the beginning Dad employed a rather sad looking elderly man who warranted the title of 'Carter'. He spent all his time looking after the two horses. After a year or so he just disappeared ... Mother seemed to think he had died. The result was that Dad took over the horses himself. It was noticeable that in jobs where both horses were needed together there was always the tendency for the outfit to move in circles unless Whinnie's speed was restrained. I recall 'Aunt' Mac saying that peasants in the far East had similar troubles when they used one horse and a water buffalo. I don't know if she explained this to Dad but I assumed that was why paddy fields often took on a half moon shape.

However Dad used to use both horses for his single furrow plough, I recall that as our fields were mostly on chalk with a top layer of 'clay and flints, when Whinnie put on a burst of speed the plough would rise and Dad would be lifted into the air before a tug on the rein and a yell of 'whoosh' would stop her and Dad - being a lightweight - would have to drag the plough back Perhaps 'Carter' had a happy release!

In due course the wheat was sown and in the spring the ground was rolled. In about September Dad would pronounce the grain ready. The first swath around the patch would be cut by hand ... using one of Dad's scythes; these had been handed down to him and through constant sharpening with a whetstone had become thin and wavy in places along the blade. A lot of sharpening went on ... as scything was tiring work it was perhaps a way of getting a break ! The swath had to be gathered and tied with three stalks of the crop ... with the end tucked in.

It was then that Dad's pride and joy .... his father's one-horse Trapper, a harvesting machine that combined artistry with simplicity. A cross between rowing and chariot racing ... with Whinnie between the shafts. It had one roller in the centre with a tiny wheel at the far end of the 'bed'. It had a large revolving fan above the 'bed' operated by a foot lever. When sufficient corn for one sheaf had been cut and laid upon the platform (the bed) at the back of the reciprocating knife, Dad had a wide angled rake to push the corn off the bed. The bed was then lifted for the next sheaf.

Every sheaf had to be gathered, butted on the ground, the thistles pulled out, and then tied with the three stalks. It was a slow, prickly and back breaking business. The tied sheaves had then to be gathered into sixes and arranged into a 'stook'. It was with this gathering of the sheaves that mother introduced we twins to real farm work. Unfortunately a few thistles remained in the sheaves and our interest was short lived. Peter decided against any form of farming at this stage.

As the field would be cut from the outside inwards and livestock - mostly rabbits -but sometimes a deer, fox, hares. pheasants or partridges - would seek the cover of the standing corn until they were forced to dash for the hedges. But as the work proceeded small boys would appear as if from nowhere and the sheaf tying would stop. All were armed with long sticks. Dad's dogs would be unleashed and Mother would keep us at a respectable distance.

All would be tense and expectant until suddenly the first victim would break cover. The idea was to make as much noise as possible to paralyse the creature with fear. The rabbits would often collide with the sheaves and in their mad dash lose direction. Men boys and even sometimes girls would seek delight in this murderous sport. It was suprising how much yelling they could do. Never until the last inch had been cut could it be certain that a rabbit was not hiding.

As the booty was collected each helper could claim a rabbit and the little boys disappeared as silently as they had arrived.. . each clutching a prize. The victims were then gutted and the dogs given a feast of intestines.

The stooks were, by tradition supposed to remain in the field and ' hear three lots of Sunday Church bells' before they were supposed to have been ready for hauling to the stack, In our case this meant loading onto a wagon and hauling to the rick yard at the farmstead.

During this period there were problems. Rooks would strip the sheaves of grain so Dad used to dash out with his twelve bore shot gun .The idea was to suspend a partly plucked carcass on a stick stuck in the ground to deter other rooks. Some farmers had scarecrows whilst others bribed small boys to 'walk' the fields.

The corn was hauled in Dad's wagon . A heavy model with two brake-shoes, metal gridded side platforms and heavy wooden lades. Flower worked in the shafts with Whinnie in the trace harness up in front. So although it would go in straight lines its progress was at times erratic.

As with haymaking the body collected a variety of creepy-crawlies and the journey to and from the field was a great treat for guests and myself. As the wagon and horses proceeded along the line of stooks Dad managed to control the stop-go of the horses with a shout of 'whoaaa!' or 'lead on!'. But as Whinnie's movements were somewhat erratic disasters could happen and more than once the entire load fell off the wagon due to her jerky movements. So it was a privileged job for a guest to hold Whinnie's halter on an extension lead.. To steady movements and to ensure that the guest did not get bitten. .

Back at the rick yard there were several old concrete 'mushrooms' lying around. In former years the corn stack was build off the ground on a frame supported by these but Dad never used them and the corn stack was built on faggots
He used to use the stack to exercise his ferrets . Especially the smaller ones. These were said to have been crossed with stoats to make them keener to chase rats.

Dad would sometimes put a small acreage of corn under cover in a shed or in the end of the barn. As a small quantity did not merit a threshing machine Dad had to invent one of his nil-cost stratagems. As he also wanted the stalks for thatching he could not use the traditional flail, although he had no doubt tried a home-made version.

However he used to cut the ear off the corn and rake the stalks through his home-made bank of nails. Thus he ended up, very laboriously, with some thatching reed, some loose bedding straw and a mixture of grain, chaff, husks and sundry bits of weed heads. This he sacked up.

Then he borrowed an old 'winnower' from a neighbouring farm.This machine was fed with the mixture and somebody would wind the handle so that air raced through and would blow out the lighter parts leaving the heavier grain and weed seeds to be sieved apart. It was a slow tedious job for wet days. If the sheaves were outside they had to be brought under cover and the stack sealed with a tarpaulin.

After the corn was in Dad often had a single line of potatoes to hand dig and collect for taking down into the cellar beneath the house. The final crop was a few acres of mangel worzels ... normally the large golden coloured specimens but usually a few smaller 'intermediate' red ones appeared. Dad also put in a few turnip and swede seeds amongst them just for 'the house'. The mangels were gathered up in the putt and thrown into the end of the barn through the end window. These jobs were , I was to discover , muddy, cold and often wet.

Chapter 19
THRESHING DAY

In winter the sun would set behind the tree-clad summit of Windwhistle . Dad would look out to the multi-coloured clouds that acted like a bright screen behind the delicate tracery of the now leafless beech trees and the pinnacled tops of the firs. If it had been a fine spell and the glories of a red sunset had bathed the countryside in a warm glow then it could be threshing on the morrow. A heap of coal and a barrel of water already graced the rick-yard.

So as the sun set we would all listen for the telltale sounds that proved that the threshing outfit was on its way. At first there was the unmistakable ringing sound of heavy iron wheels on gravel strewn tarmac. Then as the first tawny owls hooted their greeting at dusk we could make out the gentle 'chuff...chuff....chuff' as the great Steamer hauled its long load up to the final hill.

A little later a louder puffing indicated that the outfit was negotiating the old gateway onto the Purtington lane. Then an easy entrance down the long farm track to the rick-yard. Here under the knotted ash tree at the entrance Dad would await the arrival. It was always very formal.

> "Evening Driver", Dad would say .. completely ignoring the driver's mate who was already untying his bicycle secured to the back of the thresher. "Evening Varmer ", would say the driver as he deftly stopped the flywheel with a touch of the controls. "Nice zunset ... reckon it'll be vine in the marning." .
> "Looks set fine. Ready to start at nine in the morning."

The two men would then have a quick look at the position of the rick and agree where the machine would 'fix' in the morning. The driver would then uncouple the trusser and all three would pull in just off the track. The engine would then pull the thresher off the track with a burst of power. As the engine worked from the rear of the machine Dad usually suggested that the engine go to the far end of the yard to turn . This had the effect of rolling the road for free and also giving us twins an exciting chance of seeing and hearing the engine manoeuvring outside the back door. By now it was almost dark and the mate had disappeared up the road pushing his bicycle. The driver would have released his car from the tow-rope and was grinding up the track in low gear. Mother would, as a great treat, take us twins .to see the engine as it slowly went to sleep gently snorting and quietly crackling as it did so. Against the firebox would be a little bunch of lighting sticks which would be bone dry and still slightly warm when the driver returned at 6 o'clock the next morning. A wonderful smell of hot oil and steam permeated the whole area. A smell that only steam engines produce..

As a very small boy I always thought that farm work started in the last half of the night. If ever I woke in the night Mother would so often say that Dad had already got up. We would perhaps hear the lighting of the kitchen range , with the noise of buckets being moved around for use at milking .. the lynch pin of our existence.. But on the day of the threshing all farm work had to be cleared by 9 o'clock. So I never knew if work stopped at all.

The Driver would have first to hammer up the great knobs of coal and break up more sticks. Smoke would waft across the rick yard and the driver's face would light up as he opened the fire box. Dad would be seen making forays to the yard and discussing where the new straw stack was to be made and where the wagon would be placed to receive the full sacks.
Even before dawn a second figure - the Driver's mate - could be seen unloading the belt and the numerous items that were concealed in the thresher for the journey. The great moment would arrive when steam was hissing and the great beast gave its first puff of the day.
The eleven ton monster was normally such a gentle creature with its gently mumbling fly wheel and the fascinating little gyrating governor. It seemed to me to have infinite power. But as we listened it could suddenly roar as one wheel of the cumbersome dusty threshing machine was hauled up on a block of wood .Although the ground in the yard was undulating and bumpy all the machinery had to he perfectly level - and there were in-built spirit levels on the frames.

Soon the engine would be driven to its position and the great drive belt would be eased onto the flywheel. By this time the thresher side boards had been let down. The mate would be looking down at the drum from the top platform. The puff... puff puff of the engine would set the drum whirring and all the vibrating sieves started to move. After a minute the Driver would just tap the main control forward and the dust would rise from the thresher as it gained its working speed..

In those early days it seemed that on threshing day strange men would just appear at the farm.. Some from neighbouring farms and a few even followed the thresher around. Even a few small boys would appear just for the fun of killing (clubbing) the rats with sticks as they would frantically race away from the stack as it was taken down.

Everyone had their place. The farmer (Dad) was in charge of the sacks. Six hung down at the front of the thresher. A small glass eyepiece looked into the rotary sieve that sorted the grain . Four sacks were for good grain . The fifth was for 'tail corn' the small pinched grain that would be used to feed the poultry. The last sack would collect weed seed that could be fed to chicken . The Driver was in charge of the engine and the machinery and was always on hand in case of emergency. The great belt was at neck height for a small adult person. Belts along the sides were unprotected.
The Mate did the feeding of the single sheaves into the drum which was exposed in front of him through an open trap door.

Next to him stood the bond cutter, the only job that could be done by a woman, who cut the sheaf string before the Mate fed it in the drum. The sheaves were put to him (or her) by one of the two men on the rick.
At the rear of the thresher the straw - now minus the grain - fed out across the 'straw walkers' into the trusser. This delightful - if temperamental machine would spit out the straw in bundles tied with binder twine. One man would take these on a pitch fork and hand them up to a second man . The last man or, as I was to learn in later years - a boy had the incredibly dusty job of keeping the side shakers clear of chaff and fine debris. This he did with a rake or short fork (always referred to as picks in our locality). Thus nine persons made up the team.

Usually after mid day 'dinner' the rats and mice would start leaving and Mother used to shepherd us indoors to view from upstairs windows or from a distant field. The problem here was that Mother was in charge of refreshments. During breaks, in mid-morning and mid-day she had to produce gallons of hot sweet strong tea. Normally everyone brought their own food rations but Mother had to be prepared to lay on bread and cheese. At the end of the day Dad used to give away a gallon of scrumpy. to each helper.

Thus the one day of threshing would end and then started a sort of scramble in the fading light to get the grain safely stored under cover in the barn. Dad would finally pay the outside men in the cellar with the cider barrel acting as a table. Then after a quick word with the Driver and his Mate they would begin to pack up the machinery . A loud barking roar from the engine would signal that the thresher was being hauled out to the road and that the baler was ready for attachment.

Tarpaulins would cover both machines. With the mate's bicycle again behind the baler and the Mate seated in the driver's old car, the word was given to get going.

Behind, where the stack had stood, would be an empty patch of old wood faggots surrounded by heaps of dusty cavings (husks and seed heads) all under the shadow of the new straw stack.) There would be foxes ready to pick up dead rats and mice later in the night. Next morning flocks of birds would search the site shortly after dawn for dropped seeds and grain.

As darkness began to fall the outfit would be lined up on the track. The flywheel would be slowly whirring and then a single push by the Driver on the lever would set the steamer in motion. The track would become steeper so the gentle 'puff - puff' would become more insistent as they slowly disappeared into the gloom. From back at the rick yard the moon or stars could be seen glinting on the brass of the engine.

For Dad it meant that the milking and stock feeding had to be now started. The day of the threshing was, for small boys, one of great excitement, but for the farm workers their day had still a long way to go.

Chapter 20
THE POACHER

In the countryside of my early childhood the modern forms of communication were rare. Newspapers, wireless and the telephone were rarely available and even the Postman was a spasmodic visitor as he walked across fields. A telegram would be delivered from four miles distant by a elderly retired postman who had to be revived with tea and biscuits or even cider before he was fit even to start the return journey. We twins used to watch his slow recovery as he slumped on an old arm chair in the back kitchen. Mother told us that he always opened the telegrams en route so he could discuss the contents and congratulate or commiserate as appropriate.

Our telegrams were usually to tell Mother of impending visitors so it was useful to know that " Tis alright missus .. no one's passed on" as the man staggered to a seat before he fumbled in his official bag..

Local scandals and news slowly circulated by word of mouth . Someone would say, in the local dialect of course, that ' They do zay that......' or 'Aaah, us do 'eard tell that .... '. Like the whispering game of 'Consequences ', that Mother used to play with us, the stories soon assumed a life of their own and these were the building blocks of legend.

There was the living legend of the extremely ancient looking gypsy woman who we first met on a walk and from whom Mother bought some clothes pegs and .was persuaded to 'cross her hand'. I recall that as she bent over the pram and threw back her voluminous black shawl I thought that she was going to eat us ... despite her apparent lack of teeth. But Mother had heard of her gifts as a fortune-teller and was persuaded to believe that we would achieve great things and never be nasty little boys .. a very forlorn hope !

There was no countryside character more seeped in legend than the old man who was credited as being the local poacher. His exploits were apparently known to all the 'locals' but as he never seemed to get caught it could have been that the .toffs' and their minions were too stupid or too trusting to even suspect so plausible a character.

Many of the local tradesmen at this time were known by a 'title'. A farmer were known as 'Varmer', a steam engine driver as 'Driver and there was 'Landlord' at the Inn. The owners of the now fragmented larger estates were referred to as 'Squire'. Our local poacher was known as 'Pincher'. I presumed in deference to his skill.

I had learnt from an early age that there was much to be made by 'living off the land', In our case it was our own farm but Pincher had no land of his own - other than a very small kitchen garden - so he used to borrow other people's land. I somehow felt I had something in common with him for I realised that it was necessary to have a thorough understanding of the local wildlife .. in order to outwit it!.

Pincher lived in a tiny house in the woods adjoining Moor Copse over the Hill from our farm. The best time to catch sight of him was at dusk or dawn as a silent shadowy figure - with an equally shadowy lurcher dog - as he moved like a Roe Deer, always near the safety of spreading foliage or trees. He was said to be a small man although he seemed to permanently inhabit a quite enormous bulky khaki overcoat. On his head was an equally floppy cloth cap. .
Rumour said that in his coat he carried the snares, wires. knives and nets of his trade but much of the bulk was made up with the booty. Rabbits, pheasants and the odd hare, partridge and even larger birds eggs ... such as the peewit. The exception was his left pocket which was the mobile abode of his best ferret ... who could be seen peering out ready for a quick dash in a hole after a rabbit.

Pincher was adept at catching moles and it was said he carried these home under his hat, ready for skinning. His lurcher dog had the speed of a greyhound and the sagacity of a fox . He could catch prey on the run and bring 'em back alive' to his master to be dispatched and secluded in his coat.

It was with the Squire and his keepers from the Estate from which Dad had purchased our outlying farm that Pincher was in constant conflict. The Estate reared a large number of pheasants in addition to a large wild breeding population. Three keepers tended these birds.

Across the Estate were the gibbets where the keepers displayed the vermin they shot. Crows, magpies, jays, stoats , weasels. rats and the occasional owl , sparrowhawk or kestrel were swinging from the poles being slowly sun dried. There were no foxes for the Hunt occasionally met at the Great House. I expect they would have liked to see Pincher and his dog hung up there also ... but you cannot convict a man on just legend. After all Pincher was a sort of Robin Hood to the local workers.and his missus could prepare a roast pheasant dinner fit for a King

The Pheasants were not over gifted with self preservation at the best of times, and with Pincher around they could be considered suicidal. Pheasants roost about twenty feet up in the trees at night and the largest fattest birds always advertise exactly where there intend to bed down, so Pincher only had to secretly stand beneath for them to fall into his arms. Rumour said that he would circle a dim light to dazzle them, or use a smoking paper taper to make them lose balance or just spear them in the right place with a long sharp stick.

He could easily snare them almost like rabbits or use the noose at the end of a long string with a bait method. There were stories of drunk pheasants after eating Pincher's scattered grain.

I was also told the legend of the great battle ... it went like this. Some years back the local working lads on the other side of the Hill (in Lord Paulett country) decided to have a great Xmas pheasant dinner. Pincher devised a plan to get about twenty birds from the Estate on our side of the Hill (Lord Bridport country). These keepers drank in two Inns. The Windwhistle (The Wustle) and the George and Dragon (The Gaarge) at either end of the Estate nearest to where they lived.

So a group of lads rushed into The Wustle one evening shortly before Xmas and announced that a group of robbers had just left Chard to raid the Great Copse. Other local lads rushed into the Gaarge and announced that a group of robbers had just left Crewkerne for the same copse. So two groups of keepers with guns and dogs entered Great Copse at either end. Soon each assumed the other group were the robbers and firing guns in the air advanced with much commotion and shouting.

By the time that they had recognised the trick Pincher and a couple of lads had got their pheasants from another little copse and left the scene so there was a great feast. One problem worried me for some years until I learnt this tale also.

With so many birds there must have been a pile of feathers . The smell of burning feathers can pervade a house or garden for some time. Buried feathers can be dug out by foxes or dogs and seem to get everywhere on even a light breeze. Pincher had a solution.

During the winter period and particularly on Boxing Day the Foxhounds met at the Big House. The night before - after midnight - selected men (the earthstoppers0 would quietly enter into the woods and put a few spits of earth and some sticks down each fox and badger earth so the foxes could not go to ground in the morning. Pincher (known by his proper name) was a valued and practised earth stopper who knew his job. However if you happened to see him setting out at midnight he would be carrying a large sack on his back.

This contained the pheasant feathers and it is said he put some down every hole he stopped. In a few days the holes would be again dug out by their rightful owners and the woods would be struck by a snowstorm of feathers. The foxes would get the blame of course.

The result was that after dark on winter nights the Squire ordered that a chair be placed in the centre of the woods. A man with loud voice would be employed to let out a shout about every two minutes to keep the foxes away. As a small child I used to listen at my bedroom window and hear this shout from a mile away. Later a Claxon horn was used. This may have kept the foxes away but not Pincher.

When I started to work on the farm at fourteen Dad appointed me 'rabbit catcher'. He asked Pincher to instruct me in the art of snaring. ... who better! But Pincher did not entirely approve of farmers .... they had tried to catch him on occasions. Suitably bribed he certainly showed me how to prepare the pegs, stretch the wire and select the runs. I would need much practice, of course.

However Pincher did give me some bogus advice. He persuaded me to buy a small pot of aniseed oil. This to be rubbed into my hands before setting the snares. I never knew if it attracted the rabbits but I soon discovered that it attracted every stray dog and fox in the locality. So that my snares only held a skull, a bit of backbone and the white fluffy tail when I returned at dawn.

I also discovered that a couple of Pincher's mates had the 'rabbits' on the adjoining farms. After all Pincher was also loyal.

Chapter 21
FERRETING
A Days Ferreting.

The bunny rabbit was the universal animal of my childhood. You could do so many things with them ... apart from eating them. They were the stars of the picture books, their relations could be kept as pets, their fur served the clothing industry and they were at the lower end of sport for the shooting fraternity.

Rabbits on the farm were always getting killed and they never really seemed to mind. Perhaps they knew that unless so many met an untimely end the place would soon be 'wall to wall' rabbits ... and there were always more to take their place. Indeed if it were not for the rabbits all the foxes, badgers, stoats, stray cats and dogs would turn their attention to domestic stock ... not, of course, that they never strayed into the farmyard.

I was almost reared on rabbits and I often wondered where mother learnt her skills in cooking rabbits. Abroad they always employed professional cooks ... but she might have watched the cooking of armadillos, ground squirrels or sundry wombats perhaps. For although we seemed to have a different dinner each day ... they probably were all rabbit based. Indeed there was jugged rabbit, spit roast rabbit, oven roasted rabbit, rabbit and vegetable stew and another vegetable stew - with rabbit flavour. But mother's very special was rabbit in parsley sauce. Ideally she used a happy young buck rabbit that had fed on best white clover, wild thyme, selected trefoils and a smidgen of finest meadow grasses. Dad had to do the skinning and gutting and preparing them as 'oven ready'.

To us children, we knew little else but to Dad its chief merit was that it cost nothing. Rabbits to him were there for the catching. But the catching was the problem and although rabbits seemed not to mind being killed they had evolved many strategies to avoid being caught. One of which was to live underground.

LANDING GENTLY — by PAUL SMITH

Dad had no time for catching rabbits for he was always battling with the farm work or using wire, string, wood, nails or a bit of old harness to make something that cost nothing. As a child we knew nothing of organised sport. Chasing a ball around didn't produce an end product. Even shooting cost money for the cartridges. But from an early age, as soon as I showed an interest in the farm, I was allowed to accompany one of the older farm workers, Joe, on a day's ferreting. Hopefully Joe would earn his wages with a bit over and both he and I would keep a selected rabbit for the pot. The rest were collected around the farms twice a week.

The day didn't start until mid morning when Joe had done his ration of mucking out or milking. There was a lot to carry so it was rather like an exploring expedition.
Most important were Dad's two albino ferrets. A large dog ferret carried in a box with a shoulder strap. A small 'Gill' ferret could normally be carried in one of Joe's overcoat pockets. There were about a dozen green purse nets, a bill hook. Joe's pocket knife to gut the rabbits. Also a digging spade and a long sharp iron spike called a 'marker' to find a hole underground if the ferret got 'laid up', we also carried a few hessian sacks to sit on or drape over our shoulders during a storm. Joe carried sandwiches and some cold tea in a bottle. I had some biscuits and some weak lemonade.
On the 120 acre farm there would be about six miles of hedges and most of those were banked and had their own rabbit holes. Joe would usually arrange with Dad in which direction we would start. I would follow him a few paces behind usually carrying the spade and the marker.

The idea was to creep up very quietly on a few holes in a bank. Covering them with the purse nets .... so called because if the surrounding cord was attached to a stick in the hedge the speedily exiting rabbit would hit the end of the net and close it like a purse. That is, of course, shortly after the ferret was quietly placed in another hole.

The difficulty was in dealing silently with the first rabbit to get caught so that the next rabbit did not get alarmed and refuse to exit at all. A spare net would be placed over the hole and the netted rabbit extricated and dispatched with a sharp hand chop to the back of the neck.

Having cleared all, or nearly all, the rabbits at one burrow the idea was to pick up the ferret and move on. We usually moved some distance away where the rabbits were, hopefully, resting quietly.

Joe was not gifted with great patience and he would cuss the ferret which. in my innocence, I assumed would not always come to his call. It soon learnt that rabbit burrows had their main tunnels with a number of blind ended shafts which they probably used for resting. Often the rabbit would come to the netted entrance, peer outside, and find the enemy - in the shape of Joe and myself - waiting outside and licking our lips. The rabbit would then bolt back perhaps past the ferret and go head first up a side tunnel. The rabbits then expanded their haunches to fill the tunnel so that the ferret could not creep past.

This was when the ferret would get 'laid up', for ferrets were very persistent in following their prey and being unable to administer the 'coup de grace' at the neck would vainly scratch the rear of the rabbit until it was tired and then curl up and go to sleep.

Before this final state happened efforts would be made to locate the ferret and dig down to the hole. The policy here was to listen at the holes with an ear to the ground. Although my young ears were pretty sensitive I realised that the silence of the countryside was indeed anything but silent. Insects, birds, the rustling of leaves and even distant dogs barking would 'mug' the underground rustling.
Sometimes we would be lucky and the angry rabbit would stamp its feet. Joe would hear this and with a shout of "I'd 'eard un.!" demand I hand him the spiked marker and he would push this down into the soil. If it reached a tunnel this could be felt and digging start. If we were lucky we would reach down over the ferret which could then be brought out. An arm was then inserted to retrieve the luckless rabbit. I once witnessed Joe pull out four rabbits from one digging.

Another policy was to introduce the larger ferret which could sometimes shift the rabbit or at least get the first ferret to come out to complain to Joe and get caught.

If all failed we would use our second ferret to continue at another burrow. I would go back periodically to see if the first ferret was visible. Often when all was quiet it would come out to go into its box where some titbit would be placed.

Thus we would proceed across the farm with periods of digging, listening and quietly waiting. After several hours the ferrets would get tired and less keen to search out the burrows.

Joe would continue and have his snacks at the burrow but Mother preferred me to walk back to the house for dinner. Rabbit of course.

Ferreting in deep sand or chalk burrows was a bit of a lottery but there was always the chance of arriving back with a dozen or more rabbits -gutted and threaded on a carrying stick - with two sleepy and well fed ferrets and for me having seen a bit of the wildlife , the birds or beasties, as we kept quiet waiting for a rabbit to dash into our nets. A very satisfying day for a small boy.

Chapter 22
THE YOUNG FARMERS

By the time I had left school and started work on the farm Dad had reached a low ebb in his life. He had lost his old pioneering spirit and had told me that the only way to ride out the farming depression was to cut down on stock and every possible expense. It was a question of recycling old harness, machines and tools. Saving old bits of board, straightening old nails and getting nuts and bolts out of derelict machinery. Some of his scythes and hooks had been inherited from his father and had been sharpened with whetstones so many times that they had become wavy and almost paper thin. Some of the mucking-out picks had been worn down so that their prongs were almost half length.

All muck and other waste had to be saved to put on the fields. With less livestock very little bought food was needed. The poultry on free range found much of their own food around the buildings and hens could be seen chasing after flies and even catching small mice and voles. We humans lived well, with rabbits as a main dish. Hares, pigeons and surplus poultry added variety. We could, in season, always find mushrooms, nuts, sloes, crab apples and blackberries.

Mother still had her paying guests and her own car and looked to the future ... probably through the eyes of us twins. Peter stayed on at school for another year and had decided to study wireless technology by correspondence course. (At 17 Peter left permanently to make his own life starting in London) I took a short correspondence course in Commercial Art ... often drawing after dark by the light of a single candle.and trying to write a few short stories.

Dad seemed to believe that any farming skill should be gained solely through practical experience. The furthest he would go in learning from others was to lean over the neighbour's gates. There was a distinct mistrust of the new farmers who had tried to farm from a basis of dubious theory. The established farmers all seemed to know of someone who had attended a spectacular bankruptcy sale of 'They book farmers'. So Dad was perfectly satisfied if I acquired just the practical 'basics'. Knowing how to harness a horse, hand milk or calve a cow, use scythe, picks and shovel to good effect. The rest would come from practice ... hard manual work on the farm. As a labourer I would be drawn into every activity... from corn harvest to ringing pigs.

So it was Mother who jumped at the idea of my attending an inaugural meeting in Chard to start a branch of the Young Farmers Club movement. The chief instigators were, perhaps strangely, a retired butcher (Mr Gillingham) and an auctioneer (Mr Clarke) both highly respected amongst the local farmers. There would be an 'Advisory Committee' of farmers. Many of the local farmers were Methodists and their attitudes were strictly neo-Victorian, especially as far as young people were concerned.

The Club was open to both sexes from twelve to twenty. years old. but the members were all about my own age and many, especially the girls, arrived at the early meetings heavily chaperoned. With the Club Leaders and officers sitting at a table the rest would sit on chairs along the walls of a small public hall.

The basis of membership was that each member was to take possession of a calf .. about eight weeks old. A detailed record must be kept of everything it was fed and at the end of a year there was to be an exhibition where the calves were judged and sold. The 'Best Young Farmer's Calf of the Year' would hold a Challenge Cup. As the calves were Dairy Shorthorns - a dual purpose breed - some were fattened for beef and others were run on grass to become milkers. I bought my own calf back and she entered our herd after two more years.

At the regular monthly meetings of the Club the books would be inspected before any further business. At each meeting there would be a lecture or demonstration. Then we would arrange social activities .. dances, whist drives or farm visits.
But before the members could do much we had to learn the procedure of public meeting and our leaders were very insistent about such matters as addressing the 'chair' and taking a proper vote. We also had to learn to speak properly to a public meeting A local dialect was not objected to provided the meaning was clear .. with no sloppiness.

I soon overcame the shock of being elected chairman ... not very difficult under such strict guidance. The secretary was a very quiet and diligent girl whose parents saw that she did not overindulge the social side or linger after the official business.

The Young Farmers movement was a complete cure for the social ills besetting the young farming children at that time and the movement soon swept the country.

Each County Federation had its annual Rally and Clubs competed in an array of competitions from sheep shearing to tug-of-war.
One of the features of those early years was cattle judging at the markets or agricultural shows. Here four fairly similar young cows or bullocks were paraded before a Club team of about six members. Usually led by stronger girl members carrying the labels 'A', 'B', 'X', and 'Y'. These animals had to be placed in order of merit and the reasons given before a panel of expert judges.

The movement had its own national magazine 'The Young Farmer' with club news and an annual essay competition. I was lucky to have won a shared first prize at fifteen and they started using my cartoons in 1937. During those early years I cycled to local meetings and Mother drove me further afield to visit other farms and shows.

At seventeen I got my driving licence and we got a farm car. Mother and Dad were self taught and still drove in their own individual style... Mother never learnt to back properly.

As farming began to pick up, my work on our farm increased and I shared my spare time by adding a few hours each week at Yeovil Art school and in early 1939 the Territorial Army as a gunner in the West Somerset Yeomanry at Taunton.

The Young Farmers membership had gradually increased in age but was to emerge after the War to sweep the countryside as a great matrimonial agency, as the maximum age was increased to twenty five. The girl Club Secretaries suffered a huge casualty rate as they were expected to resign on marriage. I should know from first hand!!

Chapter 23
SCRUMPY CIDER

Money was not the only currency of my childhood. The village men who helped Dad from time to time were paid not only in cash but by being given the 'freedom of the cider barrel'. Farm work was always hard but so often hot and dusty. I vividly recall the scene, when the day's work was ended, when Dad joined the helpers in the cellar.

Dad would prepare to hand each man a few coins as they gently rocked on their feet, clasping a firkin of scrumpy in one hand and a very dead rabbit in the other. Their red and shining faces resembling the setting of the summer sun, For although scrumpy was sometimes said to rot the gut it was indeed a very happy drink.

Every farm had its own cider apple orchard. The names of the cider varieties it contained were often lost in the dim mists of antiquity and the orchard also contained a few small trees of 'eaters' - such as Morgan Sweets or Tom Putts - to the delight of small boys.

The orchard was a special field also used for calves, lambs, bees or 'difficult calvings'. It thus had to be near the buildings and specially guarded. There were tales of small boys receiving a thrashing from irate farmers who caught them as they descended from the branches with stomachs bloated and pockets bulging.

Our orchard was different. The farmstead buildings were in the eye of the Westerly gales that beset. the hills. The old orchard - with then blinding logic- was sited at the bottom of the combe near the water course. with an ancient woodland (Pool Copse) wrapped around its windward side. But later this wood had been interplanted with spruce and larch which reached over a hundred feet into the sky drawing up many great hardwoods with them.

This had the effect of acting as a funnel creating quite lethal circular air currents in the orchard. The old apple trees were thus gnarled and twisted in all directions and heavily swathed in feathery lichens. Several had hollow branches for owls to nest and there was mistletoe growing in their crowns. We were never troubled by small boys for although no two trees were alike their abundant splodged and mottled fruit tasted quite horrible ... like sawdust laced with cabbage water and vinegar. Dad said that our ancestors must have planted them to produce a very special blend of cider. Dad had newly planted our 'eaters' around the garden.

The ritual of the apple harvest and the making of the cider was a joyous annual event. The season started when the first real gale of autumn had blown most of the very ripe apple crop to the ground. This was followed by branch shaking, poking with long sticks and posting small boys to the outermost isolated branches so that all the fruit was grounded. without, needless to say, too great a sacrifice -or physical damage - of small boys.

By now the buildings and outhouses had been scoured for used hessian sacks. Sacks, seemed to sprout from all around the farm. They were the universal covering and towelling. They served as doormats, hand towels, overcoats for humans and animals and could, at a push be even used to strain milk. A heavy old 2 cwt grain sack that had acquired a patina of grease or engine oil was much valued as a shoulder drape over another sack apron around the waist and tied with a piece of old rope or string. Sacks were normally only washed by natural rain or even melting snow.

In a good apple year every sack had to be assembled and on being taken down to the orchard by horse and putt enjoyed an annual treat by being thrown in a heap into the small stream for the night. As a cleaning agent the water of the stream was only fairly clean at the best of times. The sacks acted as a sort of dam to filter out the fauna and flora that floated in its waters. The larger creatures- frogs, newts and sundry eels - could be shaken out when the sacks were laid out to slightly dry under the trees the next morning.

I recall that usually a few odd helpers, often older men, would turn up to help pick up the apples into buckets and tubs to be tipped into the sacks; the more grime ridden sacs were the easiest to fill as they would almost stand up by themselves

When the sacks were finally tied up they presented a slightly oozing, squishy and sticky mess. Complete with wasps, woodlice and other minor grassland fauna. Before Dad would declare the crop cleared he would inspect the trees. Invariably he would point out a few choice apples still 'smiling' from the utmost high extremities and they would wait chatting beneath as I inelegantly tried to prove Darwin's 'ancestral ape' theory. Often these bright high apples would be mere shells having been eaten out by wasps.
As the ascent of Tip Hill was too steep for a wagon, Flower - the Shire mare -

pulled the putt up the winding track with all helpers pushing at the back. As the cider press was two miles away by road in Chillington there was a small procession with Dad always in charge. While Flower would take all the sacks in the putt - which had probably been used for muck hauling the previous day - Dad would lead with Whinnie in the trap. This was piled high with 2-hogshead barrels, straw, buckets, a stone fork, a shovel, a couple of short four prong forks and a rather ancient rusty tundish (sieve) and a besum . Also needless to say, a couple of firkins of mature cider ,probably containing the strained dregs of one of the barrels to be refilled.... this would ensure a ready supply of willing helpers at the scene.

The Village cider press was at one end of an old barn used by the village carpenter and wheelwright to store wood and sawdust. The great screw hand operated press was over the collecting vat - half an old four hogshead barrel sunk into the ground beneath.

Dad was in charge during the skilled job of making the cheese - alternate layers of straw and apples. This had to be just right to ensure that on squeezing only juice seeped out and into the vat. As the apples and straw varied in consistency - the latter being bedding quality after reed combing (probably some still warm from the cows!). Dad had inherited a practised eye for this operation although he was normally tee-total. His father died - it was said- due to a surfeit of cider but the thought had just crossed my child-like mind that he could have been poisoned.

After the cheese had been repeatedly squeezed, added to and even replaced I would await the moment when the liquid in the vat would reach the lip of the vat. This would reveal, floating on top, odd splinters of wood, feathers and other sundry debris ... all liberally laced with woodlice, earwigs and other sundry beasties all struggling, like shipwrecked sailors trying to board a raft. When the top had been raked off with a flat piece of wood one of the more knowledgeable adults present would take out a small tumbler full and hold it up to the light.

"Full of body .... good colour!" he would exclaim as Dad smiled with satisfaction. I always thought that the man said 'bodies' but either could have been true. At this stage I was allowed to dip the communal tumbler. Luckily I was not asked for an opinion .. but it was wet and refreshing ... and the only drink.

The barrels were then set up in the putt and with buckets filled through the tundish. Then the dry cheese would be loaded into sacks for pigs or poultry to pick over and the slow triumphant procession go back to the farm . The barrels would then be man-handled into our (coal) cellar and secured on wooden blocks about a foot off the floor. At this stage they contained only apple juice.

Dad had probably inherited the barrels as treasured family heirlooms, for being never really cleaned out they contained the seeds for a good fermentation - other farmers sometimes used ex- rum or sherry barrels If a barrel should dry out then it had to be 'plimmed' with water before being again cider-tight.

The mysteries of fermentation were beyond my childish mind but I was told that there were long filament-like growths in a good barrel ... called 'moors'. If the barrel failed to start fermenting then it could be started with a well used sole of an old boot a piece of rancid bacon or - I was assured - a dead and partly sundried rat inserted in the bung

Dad used to keep a watchful eye on the fermentation and periodically release the internal pressure by temporarily 'easing' the bung. A further refinement in cold weather was to allow turkeys to roost on the barrels around Xmas time. I was told that barrels could explode if they were not tended carefully. Mother used to let us sometimes sleep in the small bedroom above the cellar in the winter and I had visions of a tremendous explosion underneath and my waking up to find myself floating a hundred feet in the sky drenched in cider in an assorted mixture of bed-clothes and turkey feathers.

The real old-time scrumpy made by these secret methods certainly had a great kick in its taste. The first sip would go straight through the roof of your mouth to temporarily paralyse the speech for a couple of seconds before you could utter the cry so beloved of local farmers.
    "Gaaaaaa ... that be a good drop of Zider !!"

Chapter 24
THE WILD LIFE

Windwhistle Farm was my home from infancy to old age - over 62 years- and I knew every square inch of its 120 acres, I saw both my parents die there at great age and it was where Monica - a childhood guest - joined me as a wife, after my return from WW2, and we reared four children and became great grand parents.
In the security and happiness of childhood my recollections are still fresh and vivid. There is nothing that my subconscious would wish to exclude. My twin Brother Peter must have inherited Mother's wander lust for he left the farm, in a few years, for foreign parts and never returned again except for brief weekends.

So let us go back over the years and seasons of the 1920's when Wild Life on the farm was abundant yet each small place revealed something special. From an early age I used to visit these secret places and even stray over the boundary in a circular tour. I was usually quite alone for I had watched the movements of the fox and could go cross-country through hedges often more quickly and always quietly by myself. This time I have supplied a map to make it easier for us to keep together.

Our tour starts at that once forbidden gate to Tip Hill that later gave me my freedom of the farm. Here we look down on the stream and the tall trees in the valley floor over eighty feet below. A small rookery supplies a constant cawing as the rooks ride the air currents above the trees.
Probably due to the general untidiness of the farm buildings there is a lot of insect food for small birds and the garden hedges have small birds nesting down their length - robin, wrens, and the bluetits nesting in an old stump. We come to the farm quarry and pick up a fossil. Here were get ammonites, lamp shells and the odd scallop. This is the only place on the farm where I have seen a Small Blue butterfly.

Going forward is a magnificent ancient beech tree. My first swing was an old motor tyre fastened by a chain to a lower outlying branch. I never found a bird's nest in this tree but most years it supplied a great feast of beech mast. To celebrate the Silver Jubilee of King George V in 1935 I tied a Union Jack to its topmost branch,

The field beyond was extraordinary in that it 'snowed' mushrooms and in a good season we have picked up to 20lbs in weight ,on one short early morning visit. At the hedge junction was the Owl Tree which, in my early years always had a Tawny owl nest and I learnt to blow in to my folded hands to get them to 'reply' to me at dusk. This tree was a nature Reserve in its own right... as well as a tree house!

The field beyond was, before the pigs got to work, speckled all over with ant hills. Amongst these, moles were very active with even a couple of large nursery mole-hills that had grassed over. From the steep part of this field we looked right across the Cricket Estate woods and beyond the fifteen miles to the sea at Seaton gap.
Right down by the road was a worked-out small building-stone quarry. Here I used to look for quaking grass and the small rush known as 'Good Friday Grass'.
Here I used to stray to look at the Wild Thyme and the miniature fern (Maidenhair Spleenwort) and Wall Rue and the colourful insects that visited them.

Pool Pond would often have Moorhens,Teal, and Mallard nesting. There were eels poking out of the muddy bottom - some being large silver and ready to make for the sea. Pool Copse consisted of a variety of trees. There were oaks and ash and a huge sweet chestnut but most were larch, spruce and some high black poplars. Apart from the rooks there were a variety of other birds.
Our field that bordered the Copse was on greensand and badgers had a series of setts along the bank. These badgers continually dug out fossils - mostly bivalves and marine worm casts.. This field also grew Giant Puffballs which looked like stranded sheep.
Further on we get to the old Cider Apple Orchard containing about a dozen old trees twisted and gnarled in the winds that bordered the high trees. There is one spot in the hedge where the bluebells are pure white. Horseshoe bats sometimes quartered the woodland edge at dusk.
At the end of the Orchard there is a path to the Hydraulic Ram in the Copse. A large high old ash tree carried a collection of Polypody ferns growing on the mossy surfaces twenty feet up. There is also the only patch of Wood Sanicle found on the farm.

Past this is a wet spot with springs to collect water for the ram to pump up. Also some fine marsh Horsetails, figwort and the poisonous Water Dropwort, as well as Great Willowherb and Angelica. There are rare marsh Orchids and - only once - Marsh heleborine.
Next we pass the Weir with its growth of Watercress, Golden Saxifrage and Marsh Marigold. A small Rowan Tree and a Guelder Rose were in the bank beside the

water. The Stream starts in a spring in a pot hole, usually containing tadpoles, at the top of the Weir. Various willows, sallows and fallen branches, often with a Willow Tit's nest, criss- cross the centre.

If we go to the two fields to the far side of the Weir we find a profusion of the white flowers of Earth Nut and the Callous-Rooted - Dropwort (both of which I would eat although I believe the latter is somewhat poisonous!) The hedge between these fields had a large crab apple tree. Both these fields had mushrooms on their flatter tops. These fields were good for the larger butterflies - including the Pearl Bordered Fritillary - due to the huge Bull (spear plume) Thistles flowering there.

As we near the Fairyland gate there is usually a Willow Tit singing. Marsh Tits do not come this far up the farm. (Fairyland is dealt with in another chapter). So through here to 'Lower Six Acres' at the top side a high bank has white patches. These are the badger diggings. This field has a thin soil in parts and Yellow Rattle joins the white clovers and shorter grasses like Timothy, annual meadow, quaking grass, crested foxtail and also plantains and a little red clover. In this field I also found the strange Orobance (Broomrape). The corner of the next field was a great place to see blue butterflies.

I often strayed into Eighteen Acre Copse. It was an example of ancient woodland, although many oak trees had been felled for the Great War. Here grew Herb Paris, Twayblade, Butterfly Orchid, Enchanters Nightshade, banks of Bluebells and an area of Wood Garlic. There were badgers, foxes, roe deer and a few squirrels (red at first). There was a large Rookery in the middle. Part of it was coppiced for hazel rods used for making sheep hurdles.

Our field Higher Six Acres was ploughed during the Great War and must have been reseeded with a very interesting mixture. This field had Red Poppies, Wild Mignonette, Mountain Pansies, Cudweed, Shepherds Purse, Wintercress, Charlock, Lady's Smock, Fumitory, Knapweed, Hawkbit, and well as such tiny plants as Pearlwort. And this in addition to the usual thistles, dandelions buttercups and a wide range of grasses. So we proceed to the Top Field. The soil in this field contains some rounded gravel. Indeed the flints and other stones in this field are rounded as if they had been in a fast moving river at one time in their (Cretaceous) geological history. Bracken grows in the hedgerows and there are many foxgloves. The Parish boundary fence - marking the furthest corner two acres - is very strangely made of Gorse. on a natural fault in the surface. (Dad removed this fence several years later). The cross country views from this field range from the Bristol Channel out from Burnham-on-sea to the English Channel through Seaton Gap. On one occasion (1937) when we were building a haystack in Top Field there were thunder storms around and ships were visible in the Bristol Channel as was the

Welsh Coast ,. I have climbed a tree in this field to also see the Dorset 'Knobs' to the South. Adjoining the main road (A30) there is a magnificent spreading Beech tree and near its foot on the bank was always a small patch of Footstool Calamint. The hedges also grew huge blackberries we called Luscious Berries. Honeysuckle and nut bushes also allowed a colony of Dormice. From the age of about ten I would delight in this field to view migrating birds. Fieldfares, Redwings, Gulls, Geese and the odd Raven - would appear to be flying from coast to coast.

We now go to Hilly Cross with a marl pit on its top side.. The top hedge was always sunny and I regarded this as a good place to see butterflies. In particular the Small Copper. The hedge had a lot of Tufted Vetch and Hedge Bedstraw.

The old Purtington track to Chillington - a mile in either direction - was under the far hedge and through a gap in a roadside wood called Hebers Copse ... across the main road to Hebers Lane. The trees here were Silver Pine, Beech, Hornbeam, Sycamore with a base growth of Black and Haw Thorns and Hazel, some privet over a ground cover of Wood Garlic and Ivy.

Along the road hedge were tall Silver Firs, Sycamore, Oaks and, at the junction with 11 acres a great spreading Beech. This hedge has revealed many butterflies and even the occasional Dragonfly. I used to go there to hear the beautiful song of the Tree Pipit. The hedge was dense with blackthorn and was a good place to find a summer Dormouse nest. This field often had Snipe in the winter and Peewits would nest there.
The large marl pit in the centre was often visited by Wheatears on migration.

Eleven Acres was a long field, with a straight tree lined hedge, notable as it grew a lot of Knapweed. Diagonally across the centre we reach the chalk pit under a canopy of large old Hornbeam trees in Home Field. Around this pit Cowslips flourished and there was usually a Magpies nest concealed in the 'false nests' that these trees form amongst their branches. Next door in Hind House were two smaller pits at either end. This field has grown giant mushrooms... like dinner plates ... but at long intervals. The normal mushroom also grows regularly here.

So finally we enter Ten Acre Lodge field. This field, in my childhood always had Woodlarks. Their song is probably one of the most beautiful. In the Lodge Copse across the Purtington road there was always the shimmering song of the Wood Warbler.

From the Farmstead to the Purtington road my Dad built his road. Most years I would see a Whinehat on this road .... and no where else on the farm. We have now done an almost circular tour as we walk back to the garden gate.

I have only mentioned a few of the highlights to be found on a nature walk. I should say that the entire farm was pocketed with rabbit holes and so stoats and weasels could often be seen, Crows also attacked young rabbits and there were foxes in every copse. The walk, with stops for special sightings would take me, about one hour and I was sad to see the Wild Life interest gradually decline over many years. Even now seventy years later there are pockets worth seeing and the Water Mead below Pool Copse has been declared a Site of Special Scientific Interest (SSSI).

A great feature of Windwhistle farm was its abundance of Wild Life .. both creatures and plant life. This was a time when 'traditional' methods ensured that wildlife and farming battled on equal terms and on our farm the wildlife seemed to be winning. My problem, as a small boy, was naming the species. Most species were of no economic inportance to the country dweller. So any neutral plant was just known as a weed ... with a division into girt (big) and tiddy (small). I had to rely on a small book on wild flowers that Mother bought me. This only identified a few species. The same lack of knowledge was shown to birds. Apart from the very well known species, such as herons and kingfishers ..all larger dark birds were known as rooks and the little ones as 'sparrers'. The animals were also assigned to very few groups. So, with all wildlfe, I had first to get to know them and gradually, as I was given books, to give them proper names. It was in early 1934 for instance that I was given the two volumes of T.A. Coward's comprehensive study of British Birds. So with the aid of this book I can give the ultimate name that I found to be correct.

As a small boy I was blessed with acute hearing. So that silence was almost unknown to me. Insects, especially bees, flies and grasshoppers provided a background to the song of the skylark, the woodlark and the meadow pipit amongst the grassland flowers. In the surrounding woodland the shrews and voles in the leafmould vied with the shimmering woodwarbler and the tree pipit.

Rarely was I out of range of rooks, jackdaws and the 'yaffle' . The drumming 'song' of the spotted woodpecker echoed across the valley to the lovely call of the peewit. In the woods beneath the hazel, my hearing enabled me to pick out the waking dormouse in the spring. as they wheezed into activity.

In the dusk the continuous squeaks of bats - mostly pipistrelle and long eared - echoed around the buildings and house. Further out the Greater Horseshoe bat patterned the woodland edge. High up, almost like a rook, the noctule bat, would take up its eliptical circuit. To get the loudest bat sound I would seek the leeward side of walls or trees so that the wind did not rustle about my ears.

At night there were the barks and screams of the foxes, the churring of baby badgers around their sett and the alarm grunt of the roe deer. Ubdoubtedly the most frightening wild sounds were those snarling and crashing fights by male badgers over disputed territory. But the sound that most scared Mother and us occurred like this. One winter night with a gale blowing Dad went out to check around the stock. We three sat in the living room, with three flickering candles on the mantlepiece, in front of the fire. Suddenly a piercing continuous cackling strident note sounded at the kitchen door. Mother gathered us to her ... not knowing what to do. The sound was not human ... nor animal .. we had heard nothing like it before. Then from the back kitchen we heard Dad coming in and Mother shouted ..Lan! Lan! Dad rushed in but as soon as he touched the door the noise stopped. His lantern had gone out in the gale and his matches being-like himself, soaked he had come in for a new box. But first he went out into the kitchen with a candle and even opened the outer door a fraction shielding the flame with his hand. He could find nothing - relit his lantern and made his way out again. As he shut the outer door the noise restarted ... this time even louder. Mother bravely left us cowering on the settee and investigated herself. But as soon as she touched the door the noise stopped as if we were being watched. Finally when Dad returned he waited with the poker at the ready just inside the door. When the noise restarted he was there with it . What had happened was that - as the house was very far from draught proof - a constant movement of air was moving though the room and a slither of old wallpaper had become dislodged from the wall and was acting as a reed inside a drafty slit between the door jamb and the wall.. Dad pulled the paper right off and held it up so that Mother could get us quickly to bed and to sleep.

Around the farm there were other strange noises from time to time. Branches rubbing together in the woods and roof timbers moving as the sun warmed tin or even slate roofs so that the timbers clicked. Sometimes the winding wooden stairs would mumble to themselves in the stillness of the longest nights.

Gradually over many years I was able to build up a wildlife map of the farm. From the top fields on the chalk up to nearly 800 feet above sea level to the wet floor of the valley with a stream down to nearly 500 feet.

There were fifteen fields and some six miles of boundary hedges and fences. There were five chalk pits, once used for chalk or marling, in various states of levelling. There was about three acres of woodland, oaks, ash , fir and larch. Holly trees dotted the hedges and numerous trees grew in the hedges because the old hedgers thought it unlucky to cut a standard tree.

Chapter 25
WAR LOOMS

It is lucky that we cannot know what our future will hold. We can only make an informed guess and take what action we think necessary. My twin brother Peter had left the farm for good at 17 in 1937. He never liked the rural life and after taking a correspondence course on electronics went to London.

It was a day of sadness when he left by train from Crewkerne Station. He had never helped on the farm. I vividly recall that evening Mother sitting down at the Piano and playing, from her book, a few movements from the Beethoven's sonatas and finally what I still think as a solemn piece, Salut D'Amour by Elgar. She hugged me afterwards and we both had tears in our eyes.

I had the dual influence of Mother who encouraged me to look outside the farm to the Young Farmers Club, of which I was now chairman and to my hobbies of drawing and writing. I had taken a course on art and I had won a writing competition and had some cartoons published. I even attended the Life Class at Yeovil School of Art for a mid-day session each week after the Principal had called on Mother and me at the farm.

Dad, although he nearly got called up when shortage made them grab any man in World War, I, was only interested in my continuing as a farm labourer and did not approve of this deflection of my efforts; but I had inherited Mothe'rs restless spirit as well as Dad's deep love of the farm. I was indeed a bit of a schizophrenic. Mother, who was in Europe for most of the Great War had cowered in a cellar in Paris when 'Big Bertha' was firing. She had lost a French boyfriend early on the battlefield and had a horror of the brutality and futility of Wars. But, with her blessing, and on the advice of a farmer friend, I joined the West Somerset Yeomanry who had field guns. I would thus avoid being called up in the Infantry, given six weeks drilling with a rifle and being sent to the trenches. Our hope was that the

War clouds would pass but I had now 'burnt my boats.'

The Yeomanry were kind to me although I was by nature a loner and not adaptable to regimentation. They paid my car expenses to Taunton for evening 'parades' and paid me to take a two week holiday in a camp on Dartmoor. I had to give up my Art classes. I would have been horrified at the time if I had known that my action would land me nearly seven years as a soldier.

The War clouds began to gather in early 1937 and farming was beginning to pick up from the earlier depression but Dad, after his efforts to get the farm going on so little capital had worked himself to a frazzle.

He had not paid me wages but by now I was the farm rabbit catcher - going 50/50 on the takings - I caught moles and some stoats and weasels for their skins. I also shared with Mother some of my takings from picking mushrooms, nuts, sloes and crab apples. I also had reared a Young Farmer's calf to maturity and had one milking cow. I had a small pen of hens and sold eggs.

In fact I worked all the hours of daylight and often beyond. Dad and I could not really keep up with the maintenance work on buildings, fences and implements but Dad would not employ more labour.

Mother, on the other hand, under primitive conditions was never short of paying guests. But she too did not employ help in the house. But she, like Dad, was nearing sixty, and was tired. Luckily I was so busy so as not to worry but just took each day as it came.

During the final summer months I had done two drills a week. My farmer friend had become an Officer in the Artillery Battery and I had managed to take his place and was training as an artillery surveyor. This did not interfere much with my farm work for I had to get up before dawn to collect my snared rabbits before doing the hand milking just after dawn with Dad.

At the end of August 1939 Mother had some of our relations as guests. Despite the foreboding it shocked us all when the final blow came. All Territorial and Reserve forces we asked to report immediately to their Units. I was working around the buildings with the wireless in the window when Neville Chamberlain made his fateful announcement. Despite the warnings the final fact of War was a shock. I came in from the farm put on my new battledress, gathered up my kit and a guest would take me by car to Taunton. Although Mother outwardly kept a stiff upper lip I could feel her trembling as she hugged me goodbye.

What happened then I have related in my book "Wars are wot you make 'em.

Landing Gently 148 by Paul Smith